CHEMISTRY RESEARCH AND APPLICATIONS

VOLATILE ORGANIC COMPOUNDS

EMISSION, POLLUTION AND CONTROL

CHEMISTRY RESEARCH AND APPLICATIONS

Additional books in this series can be found on Nova's website under the Series tab.

Additional e-books in this series can be found on Nova's website under the e-book tab.

CHEMISTRY RESEARCH AND APPLICATIONS

VOLATILE ORGANIC COMPOUNDS

EMISSION, POLLUTION AND CONTROL

KHALED CHETEHOUNA
EDITOR

New York

Copyright © 2014 by Nova Science Publishers, Inc.

All rights reserved. No part of this book may be reproduced, stored in a retrieval system or transmitted in any form or by any means: electronic, electrostatic, magnetic, tape, mechanical photocopying, recording or otherwise without the written permission of the Publisher.

For permission to use material from this book please contact us:
Telephone 631-231-7269; Fax 631-231-8175
Web Site: http://www.novapublishers.com

NOTICE TO THE READER

The Publisher has taken reasonable care in the preparation of this book, but makes no expressed or implied warranty of any kind and assumes no responsibility for any errors or omissions. No liability is assumed for incidental or consequential damages in connection with or arising out of information contained in this book. The Publisher shall not be liable for any special, consequential, or exemplary damages resulting, in whole or in part, from the readers' use of, or reliance upon, this material. Any parts of this book based on government reports are so indicated and copyright is claimed for those parts to the extent applicable to compilations of such works.

Independent verification should be sought for any data, advice or recommendations contained in this book. In addition, no responsibility is assumed by the publisher for any injury and/or damage to persons or property arising from any methods, products, instructions, ideas or otherwise contained in this publication.

This publication is designed to provide accurate and authoritative information with regard to the subject matter covered herein. It is sold with the clear understanding that the Publisher is not engaged in rendering legal or any other professional services. If legal or any other expert assistance is required, the services of a competent person should be sought. FROM A DECLARATION OF PARTICIPANTS JOINTLY ADOPTED BY A COMMITTEE OF THE AMERICAN BAR ASSOCIATION AND A COMMITTEE OF PUBLISHERS.

Additional color graphics may be available in the e-book version of this book.

Library of Congress Cataloging-in-Publication Data

Volatile organic compounds (Nova Science Publishers : 2014)
 Volatile organic compounds : emission, pollution and control / [edited by] Khaled Chetehouna
(Bourges Higher School of Engineering, Bourges, France).
    pages cm. -- (Chemistry research and applications)
 Includes bibliographical references and index.
 ISBN: 978-1-63117-862-7 (hardcover)
 1. Volatile organic compounds. 2. Pollution. 3. Pollution prevention. I. Chetehouna, Khaled, editor of compilation. II. Title.
 TD885.5.O74V656 2014
 628.5'3--dc23
                        2014014983

Published by Nova Science Publishers, Inc. † New York

CONTENTS

Preface		**vii**
Chapter 1	Physical Modelling of Biogenic VOCs Emission and Dispersion in a Forest Stand *S. Aubrun and B. Leitl*	**1**
Chapter 2	Estimation of VOCs Emissions during the Wildland Fires from 1995 to 2009 in Corsica *T. Barboni, P.A. Santoni and F. Bosseur*	**27**
Chapter 3	Biogenic Volatile Organic Compounds Emissions of Heated Mediterranean Vegetal Species *K. Chetehouna, L. Courty, L. Lemée, F. Bey and J. P. Garo*	**41**
Chapter 4	Contribution of Biogenic Volatile Organic Compounds to Tropospheric Ozone Formation in the Pearl River Delta Region of China *K. Cheung and H. Guo*	**79**
Chapter 5	Natural Organic Compounds from the Urban Forest of the Metropolitan Region, Chile: Impact on Air Quality *M. Préndez, K. Corada and J. Morales*	**103**
Chapter 6	Latest Results on the Catalytic Oxidation of Light Alkanes, As Probe VOC Molecules, over Ru-Based Catalysts: Effects of Physicochemical Properties on the Catalytic Performances *Hongjing Wu, L. F. Liotta and A. Giroir-Fendler*	**143**

Chapter 7	Removal of Volatile Organic Compounds (VOCs) Using Adsorption Process onto Natural Clays *H. Zaitan and H. Valdés*	**169**
Index		**209**

PREFACE

Volatile Organic Compounds (VOCs) have anthropogenic and biogenic origins. At earth scale, the natural sources represent a great part of the total VOCs present in the atmosphere but in industrialized regions, anthropogenic ones become majority due to the various human activities related mainly to chemical industries (liquid fuels, solvents, thinners, detergents, degreasers, cleaners and lubricants). Almost all VOCs have effects on human health and many of them are even carcinogenic. It is also known that the VOCs can affect the central nervous system and may have mutagenic effects. Apart from human health, they also play an important role towards the environment, especially in the atmospheric pollution processes. Indeed, VOCs emissions lead to the promotion of photochemical reactions in the atmosphere (ozone formation, depletion of the stratospheric ozone layer and formation of photochemical smog). The present book gathers and presents some current research from across the world conducted by scientific experts in their fields. In seven valuable contributions, it deals with the emission and the environmental impact as well as the control of the Volatile Organic Compounds.

Chapter 1 – The turbulent dispersion process responsible of the VOCs transport within and above the canopy is still under study. This issue is of great interest since VOCs contribute to the global chemical reactions encountered into the troposphere. At the forest scale, VOCs are expected to travel within and above the forest, interacting with each other or with other chemical compounds. Some quantitative information can be given about the VOCs transport mechanisms in forest through the physical modeling in wind tunnel. Indeed the forest can be replicated at a reduced scale with an aerodynamical "drag-porosity concept" and the VOCs emissions can be modeled through the emission of a passive tracer through an area source. This approach will be illustrated through the Emission and Chemical transformation of Organic compounds (ECHO) project.

The concept of this project was to combine field experiments, laboratory experiments investigating emission and uptake of trace compounds by the plants, and modelling experiments simulating the chemistry of biogenic trace gases and the dynamics of a forest stand under well-defined conditions. The chosen site was the forest area surrounding the Forschungszentrum Juelich (Juelich Research Centre, Germany). In order to simulate the dynamical properties, the forest area was modelled to a scale of 1:300 and studied in the large boundary layer wind tunnel at the Meteorological Institute of Hamburg University. The model of the forest must reproduce the resistance to the wind generated by this porous environment. Rings of metallic mesh were used to represent the trees following preliminary tests to find an arrangement of these rings that provided the appropriate aerodynamic characteristics of a forest. The turbulence properties of the flow were measured in the wind tunnel within and above the canopy. Subsequently, they were compared with field data obtained at the Juelich Research Centre, in order to test the quality of the modelling concept. The comparison showed a good agreement and results were consistent with previous studies. Tracer-gas experiments were carried out in the field within the canopy, which were then replicated in the wind tunnel. The order of magnitude of the dimensionless concentration downwind of the point source was in agreement with the field experiments.

Wind tunnel footprint experiments gave quantitative information about the VOCs origin and their transit time within the forest before that they were sampled at a specific location.

Chapter 2 – The countries of the Mediterranean basin are particularly affected by fires, which travel several thousand hectares per year. France is particularly vulnerable to wildland fires and mainly Corsica, where 84,000 hectares of maquis and forest have been burnt from 1995 to 2009. This paper present an estimation of Volatile Organic Compounds (VOCs) emitted in Corsica during this period. The authors first deal with the identification and the quantification of the main VOCs found in wildfire smoke. These results were obtained on a combustion chamber of 0.4 m^3 containing fuel submitted to an epiradiator and a sampling pump with an adsorbent tube (Tenax TA). The analysis is carried out with an Automated Thermal Desorption (ATD) coupled with a Gas Chromatography (GC) and with two detectors: Mass Spectrometry (MS) and Flame Ionization Detector (FID). 71 VOCs were identified. These emissions represent 25.4 kg/ha burnt. An extrapolation of these results for fires in Corsica leads to almost 1.92 Gigagrams of VOCs emitted into the atmosphere from 1995 to 2009. Forest fires increase the potential risk for the population long time exposure.

Chapter 3 – The effect of temperature on Biogenic Volatile Organic Compounds (BVOCs) emissions from five vegetal species was studied between 70 and 180 °C, range during which vegetation produces high amount of these gases. Emissions were investigated at small and middle scales. For a given species, if emitted BVOCs content does not change, the relative percentage of each volatile compound varies according to temperature. High amount of terpenoid compounds were emitted, except for Cistus albidus, and emissions are increasing with temperature. The main identified compounds are thymol, 1-fenchone, α-pinene, 3-hexen-1-ol and limonene for respectively Thymus vulgaris, Lavandula stœchas, Rosmarinus officinalis, Cistus albidus and Pinus pinea. The results obtained will permit to develop a database of BVOCs emissions at elevated temperature and to include their combustion in physical forest fires models.

Chapter 4 – The Pearl River Delta (PRD) of China— encompasses 0.41% of China's land area but accounts for about 9% of China's national GDP— has been a rapidly developing economic region since 1980s. Like many other megacities in the world, it suffers from serious air pollution problems, particularly ozone (O_3) pollution. O_3, produced from a series of chemical reactions in the presence of nitrogen oxides (NOX), volatile organic compounds (VOCs) and sunlight, is harmful to both human health and environment. As a major precursor of tropospheric O_3, VOCs originate from both anthropogenic and biogenic sources. Biogenic VOCs (BVOCs), primarily composed of isoprene and monoterpenes, are emitted naturally in substantial quantities from certain types of terrestrial vegetation. The atmospheric reactivities of most BVOCs are higher than those of many anthropogenic VOCs (AVOCs), and thus, they are believed to play an important role in the formation of tropospheric O_3. In metropolitan cities, the potential of BVOCs to form O_3 is amplified by high concentrations of NOx. Therefore, characterization of BVOCs is essential for understanding O_3 chemistry in urban areas, and for the regional air quality modelling. The following book chapter will summarize the findings from various air quality studies conducted in the PRD region, and provide a comprehensive review of the contribution of BVOCs to O_3 formation in a wide range of space and time. By thoroughly evaluating the implications of photochemical oxidation of BVOCs, more effective air quality regulations could be developed to control O_3 pollution.

Chapter 5 – Tree species emit oxygen and Biogenic Volatile Organic compounds (BVOCs) which react in the atmosphere generating other chemical species including ozone (O_3). At the same time, trees capture particulate matter and gases (carbon dioxide, nitrogen oxides, and ozone).

Ozone is a secondary pollutant especially abundant in urban atmospheres. Santiago, Chile, is affected by high concentrations of O_3, especially in the northeast of the city and during the austral summer. Due to the aesthetic, climatic, and ecological benefits derived from trees, the Chilean government has been using them in a natural decontamination strategy especially for removing particulate matter and some gases not to mention the potential production of ozone through the emission of BVOCs.

Since BVOCs emissions are species-specific, their contribution to the photochemical reactivity in urban environment is very much related to plant biodiversity in the urban forest.

In this chapter the authors report a reanalysis and summary of the different works made by our group, focus on determining EF of isoprene and monoterpenes from around 33% of the Santiago urban forest. This study also includes the ability of urban forest species to generate O_3 through an index called the Photochemical Ozone Creation Index (POCI).

This report looks at 15 trees, exotic and native, studied at different stages of growth (small, young and adult) and seasons (austral autumn and spring). Standard EF are reported according to normalization proposed by Guenther et al., 1995. A discussion considering the standard and non standard EF is also included. Results show that, in general, exotic trees are more pollutant than native trees because the EF and POCI values of exotic trees are higher than those of native species. Those results are relevant if trees are used for decontamination purposes.

Modifications made to the emission inventory of BVOCs replacing EF, included by default in the model by the experimental EF, demonstrate that the taxonomical approaches of EF overestimate biogenic emissions. The integration of the different parameters contributes to the discussion for selecting the species more beneficial for the Metropolitan Region of Chile from the environmental and human health point of view.

Also some information is given about BVOCs emissions from other Latin American countries.

Chapter 6 – Volatile Organic Compounds (VOCs) are recognized as major responsible for the increase in global air pollution due to their contribution to ozone and photochemical smog. Currently, the most active catalysts for VOCs oxidation are based on noble and transition metals.

Among noble metals, platinum, palladium and rhodium exhibit high activity and selectivity at low temperature, but they are unstable in the presence of chloride compounds.

Systems based on transition metal oxides (such as V_2O_5, MnO_2, Co_3O_4-based oxides) are generally less active than noble metals, some formulations are stable to chlorines and their cost is significantly lower.

Supported ruthenium catalysts have received much attention over the past years, because of their high activity in oxidation as well as reduction reactions. Such catalysts have been proved to be among the best catalytic systems for oxidation of various substrates, such as carbon monoxide, ammonia, hydrogen, alcohols, diesel soot, and even in low temperature oxidation of HCl. However, few studies on VOCs oxidation over supported Ru have been conducted so far, especially in light alkane combustion. The physicochemical properties of Ru catalysts have been found to strongly influence activity and stability. Therefore, the comparison between catalysts prepared by different methods and pre-treated in different conditions is difficult.

In order to give a general overview on the state of art, the present chapter focuses on the latest results on the catalytic performance of Ru catalysts for light alkanes oxidation with special attention to the structure-activity relationship.

Chapter 7 – Volatile Organic Compounds (VOCs) are critical toxic substances that may cause harmful effects on human health when are emitted into the environment. The control of the emissions of VOCs into the atmosphere is one of the major environmental problems nowadays. Many conventional methods have been developed for industrial gaseous waste treatment but adsorption of contaminants onto adsorbents and their subsequent desorption for reuse or destruction has acquired high approval. Adsorption has been shown to be a process cost-effective and environmental friendly compared to other technologies such as absorption, biofiltration, or thermal catalysis.

This chapter summarises the general backgrounds described in the literature related to the control of the emissions of VOCs. Firstly, a state of the art of the sources of emissions of VOCs, and their main effects to human health and environment is presented. Then, the main technologies for VOCs control, their principles, limitations, applications and their remediation costs are briefly reviewed. Finally, the use of natural adsorbents as an economic alternative for the abatement of VOCs is discussed.

Dr. Khaled Chetehouna
Bourges Higher School of Engineering
PRISME Laboratory UPRES EA 4229
Research Team: Combustion and Explosion
88 Bd Lahitolle, 18020 Bourges, France
Tel: +33 248 484 065
khaled.chetehouna@ensi-bourges.fr

In: Volatile Organic Compounds
Editor: Khaled Chetehouna

ISBN: 978-1-63117-862-7
© 2014 Nova Science Publishers, Inc.

Chapter 1

PHYSICAL MODELLING OF BIOGENIC VOCS EMISSION AND DISPERSION IN A FOREST STAND

S. Aubrun[1] and B. Leitl[2]*

[1]Univ. Orléans, INSA-CVL, PRISME, EA 4229, Orléans, France
[2]University of Hamburg, Meteorological Institute, Hamburg, Germany

ABSTRACT

The turbulent dispersion process responsible of the VOCs transport within and above the canopy is still under study. This issue is of great interest since VOCs contribute to the global chemical reactions encountered into the troposphere. At the forest scale, VOCs are expected to travel within and above the forest, interacting with each other or with other chemical compounds. Some quantitative information can be given about the VOCs transport mechanisms in forest through the physical modeling in wind tunnel. Indeed the forest can be replicated at a reduced scale with an aerodynamical "drag-porosity concept" and the VOCs emissions can be modeled through the emission of a passive tracer through an area source. This approach will be illustrated through the Emission and Chemical transformation of Organic compounds (ECHO) project.

The concept of this project was to combine field experiments, laboratory experiments investigating emission and uptake of trace

---

* Corresponding author: Email: sandrine.aubrun@univ-orleans.fr.

compounds by the plants, and modelling experiments simulating the chemistry of biogenic trace gases and the dynamics of a forest stand under well-defined conditions. The chosen site was the forest area surrounding the Forschungszentrum Juelich (Juelich Research Centre, Germany). In order to simulate the dynamical properties, the forest area was modelled to a scale of 1:300 and studied in the large boundary layer wind tunnel at the Meteorological Institute of Hamburg University. The model of the forest must reproduce the resistance to the wind generated by this porous environment. Rings of metallic mesh were used to represent the trees following preliminary tests to find an arrangement of these rings that provided the appropriate aerodynamic characteristics for a forest. The turbulence properties of the flow were measured in the wind tunnel within and above the canopy. Subsequently, they were compared with field data obtained at the Juelich Research Centre, in order to test the quality of the modelling concept. The comparison showed a good agreement and results were consistent with previous studies. Tracer-gas experiments were carried out in the field within the canopy, which were then replicated in the wind tunnel. The order of magnitude of the dimensionless concentration downwind of the point source was in agreement with the field experiments.

Wind tunnel footprint experiments gave quantitative information about the VOCs origin and their transit time within the forest before that they were sampled at a specific location.

Keywords: Wind tunnel, turbulence, forest, porous media, shear layer, footprint experiments

INTRODUCTION

In the framework of the ECHO (Emission and Chemical transformation of Organic compounds) project, the transport of the biogenic volatile organic compounds (BVOC) inside and above a forest canopy was investigated. The goal was to determine the net source of reactive trace compounds supplied by a mixed forest stand. The emission rates of BVOC from a mixed forest stand needed to be quantified, the amount of primary emitted VOC, which are transported directly into the planetary boundary layer and the amount of VOC, which are chemically processed within the canopy needed to be determined. Specialists with different competences, as biologists, chemists and meteorologists, were federated in that project to study these different questions. Methods of investigation used all available tools, to include field campaigns, laboratory experiments, numerical and physical modelling. The latter tool consists in measuring in an atmospheric boundary layer wind tunnel the turbulent flow and associated fluxes which develop on a forest area modeled at a reduced geometric scale. In opposition to field experiments, the possibility to control boundary conditions during testing ensures the statistical representativeness of the obtained results. Providing the Reynolds independency, it is straightforward to prove that the results about dispersion process obtained in wind tunnel can be up-scaled at field scale.

The contribution to the ECHO project of the University of Hamburg was to design a model of the finite forest area and to study it in its atmospheric boundary layer wind tunnel WOTAN. The flow properties and the transport of emissions within and above the canopy were studied. The chosen field site was the finite mixed forest area surrounding the Research Centre of Jülich (Germany). In order to reproduce the resistance to the wind generated by the trees while enabling an easy reproducibility of the model, an arrangement of opened rings made from metallic mesh was chosen. A comparable design was used by Beger (1983) to model a forest area but no further publications related to this set-up are available. In this study, a quadratic arrangement of metallic mesh was used. Hall et al. (1999) used some closed rings of metallic mesh to model high vegetation inside courtyards. Their choice was intuitive and the quality of the physical modelling was not tested. The set-up chosen in the present chapter is based on the same geometry but is not designed with respect to shape, drag coefficient or Leaf Area Index (LAI) similarities between field and model. The strategy was to achieve the same aerodynamic properties of the inside and above canopy flow, as measured at the field site.

The description of the model and the validity of the modeling concept against field data will be first detailed, through velocity and concentration measurements. After this validation stage, some footprint experiments will be presented in order to give information about the transit path and time of BVOC within and above the forest stand. Finally, the turbulent fluxes of BVOC above the forest will be quantified and some turbulent patterns responsible of these fluxes will be described.

The present chapter is a synthesis of Aubrun and Leitl (2004), Aubrun et al. (2003, 2005).

THE FOREST SITE AND THE FIELD EXPERIMENTS

The deciduous forest stand in Jülich is representative of a typical European forest area (Figure 1). It covers 350 ha and is located in a farmland-type region. The inhomogeneity in the distribution of tree species, of tree age and of tree height is significant. As a consequence, a careful tree inventory was carried out. The meteorological conditions are permanently recorded at the meteorological tower at 7 stations located from 10 to 120 meters above ground. In the framework of the research project ECHO, three additional measurement towers were built-up inside the forest in order to measure the meteorological conditions as well as the biogenic VOC concentrations inside and immediately above the forest canopy. The main tower is located in an area planted with oak and beech trees, which are 150 years old. The local average tree height is between 25 and 30 m and the local Leaf-Area-Index is 3.6, which is considered as representative of a dense canopy.

The 10-min-average horizontal velocity and its standard deviation were measured at the main tower with Ultra-Sonic Anemometers-Thermometers during 2 hours on the 13th of September 2000 at 5, 10, 17 and 30 m with a mean wind direction of 281° and a mean wind speed of 4.2 m s^{-1} measured at 30 m above ground. At the meteorological tower, the mean temperature gradient was determined between 20m and 120m as -1.59 °C/100m. According to the terminology of the VDI-guideline 3782/1 (2001), the diffusion class during this experiment was 'C' in terms of Pasquill's classification scheme (Pasquill, 1974), denoted as 'neutral'. The forest fetch of the main tower for a westerly wind direction is 1400 m with the last 370 m covered with 25-30 m trees. These measurements are used to perform the comparison with the wind tunnel data.

In September 2000 and June 2001, several tracer experiments were carried out at the ECHO site. The aim was to provide basic information about the dispersion process inside the forest and to supply a data set, to help design and validate the wind tunnel model of the ECHO site. Westerly winds correspond to one of the most frequent situations at the field site. As a consequence, the experiments were focused on these wind directions.

Sulphur hexafluoride (SF6) was used as tracer. The advantages of SF6 are numerous in that it is inert, non-toxic, has a very low atmospheric background concentration (~ 5ppt), and can be easily detected by an online gas chromatography technique. Furthermore, SF6 is a man-made gas, which is not released through natural processes.

SF6 is released directly from gas cylinders provided by the manufacturer. The cylinders contain SF6 in liquid phase. The vapour pressure above the liquid phase keeps constant at about 13 bars until the cylinder is empty. Constant release rates (~0.2 g s-1) are adjusted by a pressure reduction valve combined with a flow resistor. The achieved accuracy of the source flow rate

Physical Modelling of Biogenic VOCs Emission and Dispersion ... 5

is better than 5%. Air samples were taken by filling sampling bags (aluminium-coated plastic bags Linde Plastigas®) and were analysed off-line using a gas chromatograph (Siemens Sichromat 3) equipped with an ECD. Details of the tracer technique are described in Möllmann-Coers et al. (2002).

Figure 1. Aerial view of the Juelich Research Centre and the surrounding forest.

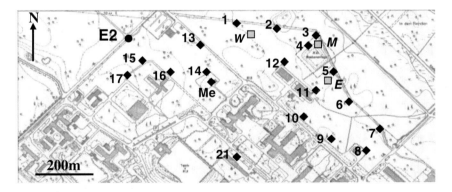

Figure 2. Experimental site of the tracer-gas experiments. E2 is the SF6 point source location, W shows the location of the west sub-site; M the main sub-site, E the east sub-site and Me the meteorological tower. Black diamonds numbered from 1 to 21 show the sampling locations.

The preconditions for the experiments were a uniform moderate wind speed (~ 3 m s-1), constant wind direction, and neutral thermal stratification.

These conditions are necessary to achieve results that are comparable with wind tunnel studies. Two tracer experiments were carried out using the tracer release position E2 (Figure 3), which is located exactly on the west side of the ECHO site. The release height is 4m above ground. In the first experiment (13/09/2000) the tracer release started at 11:30 CEST (Central European Summer Time (UTC + 2 h)) and stopped at 13:30 CEST. From 12:00 to 13:30 CEST, nine 10-min samples were taken consecutively at each position shown in Figure 2, about 1m above ground. The mean SF6 concentrations were obtained by averaging the concentrations sampled from 12:10 to 13:30 CEST. The mean wind speed during the sampling period was 4.3m s-1 and the mean wind direction was 280.2°. At the meteorological tower, the mean temperature gradient was determined between 20m and 120m as −1.58 °C/100m. The diffusion class during this experiment was 'C' in terms of Pasquill's classification scheme (Pasquill, 1974), denoted as 'neutral'. Simultaneously, 10-min-average horizontal wind speeds and the corresponding standard deviations were measured at the 17m-mast with Ultra-Sonic Anemometers – Thermometers (USAT METEK, USA-1) during the 2-hour experiments at 5, 10, and 17m.

During the second experiment (12/06/2001), the tracer release, as well as the air sampling, started at 11:00 CEST and stopped at 12:15 CEST. Three 5-min samples were taken followed by six 10-min samples. The mean SF6 concentrations were obtained by averaging the concentrations sampled from 11:10 to 12:15 CEST. The mean wind speed was 3.1m s-1 and the mean wind direction was 274.6°. The wind direction was around 290° during the first 50 minutes and shifted back to 255° at the end of the sampling period. Consequently, the dispersion process and the plume axis were controlled by the 290° period. The mean temperature gradient was −1.67 °C/100m. The diffusion class during this experiment was 'B' in terms of Pasquill's classification scheme (Pasquill, 1974), denoted as 'unstable'. Recent papers based on field experiments showed that no strong relationship could be found between unstable conditions and the turbulence structure inside a canopy (Brunet and Irvine, 2000, Villani et al. 2003). Rannik et al. (2003) confirmed this result since they showed that the dispersion process inside the canopy was only slightly influenced by unstable conditions. Consequently, the second field experiment is maintained for the comparison with the wind tunnel data, obtained under neutral conditions.

THE MODELLED FOREST SITE AND THE WIND TUNNEL EXPERIMENTS

The modelling of atmospheric flows in wind tunnels is a powerful tool, which has already been used for several decades in environmental research (Snyder, 1981, Plate, 1982, VDI 3783/12, 2000). Furthermore, the physical modelling of forest canopies have already been performed, generally in order to study the dispersion process of anthropogenic or biogenic emissions, or the constraints applied to the trees due to wind forces (Marshall et al., 2002, Novak et al. 2000, Finnigan and Shaw, 2000).

Snyder (1981) and the German Engineering Association VDI (3783/12, 2000) provide extensive guidelines on modelling of atmospheric flows in wind tunnels. The major requirements are the following: The classical strategy is to reduce all the geometric lengths (buildings and vegetation dimensions, boundary layer thickness...) to a smaller scale and to have similar dimensionless parameters describing the fluid, flow and thermal properties, i.e., the Prandtl (Pr), Eckert (Ec), Reynolds (Re), Rossby (Ro), Richardson (Ri) numbers, between the full scale and the wind tunnel situations. Since the fluid in both cases is air and the flow velocity is low (>10m s-1), the Prandtl and Eckert numbers are similar. The Rossby criterion, which represents the influence of the Coriolis forces on the flow, does not need to be respected, if the longest dimension of the modelled area is smaller than approx. 5km. No thermal effects are taken into account in the current model (Ri =0). This corresponds to neutral stability conditions in full scale. It is not necessary to respect the Reynolds number similarity (in full scale, with the tree height), if the modelling concept enables the Reynolds number independence to be reached (very turbulent approach flows, rough surfaces, geometric scale larger than 1:1000). This very fortunate property enables reference velocities to be similar to the full scale ones, e.g., <10 m s-1. Nevertheless, the Reynolds number independence was systematically checked, repeating measurements of flow and passive tracer-gas dispersion properties over the range of velocities available in the wind tunnel (in the model scale).

A model of the Juelich Research Centre and the surrounding forest was set up at a scale of 1:300 and for a wind direction of 270°. The complete model was 10.5m long and 4m wide (corresponding to 3150m and 1200m, respectively, at full scale). With a geometric scale of 1:300, the upwind edges of the forest area are included in the model. Consequently, the modelled forest fetch is equivalent to that in full scale. Buildings were made out of rough

Styrofoam. Geometric details of buildings bigger than 0.5m are reproduced in the model. By processing aerial views of the ECHO site taken in 1998, the cartography of the tree height distribution was determined and 6 different ranges of tree height were defined: range 1 $Ht < 10$m, range 2 10m $< Ht <$ 15m, range 3 15m $< Ht < 20$m, range 4 20m $< Ht < 25$m, range 5 $25 < Ht < 30$ and range 6 30m $< Ht$. This distribution was reproduced in the model. The areas of range 1 ($Ht < 10$m) are homogeneously covered in the model by tangled plastic swarf. This tangled material had a thickness less than 33 mm (corresponding to 10m at full scale). For the other ranges, a specific arrangement of metallic mesh rings was used to simulate the trees. The metallic mesh was made from steel wires (diameter 0.4 mm) with a mesh size of 2.8 mm. The height of the rings was directly related to the middle height of the modelled range (i.e., for range 5, the average ring height of 92 mm in the model corresponds to the average tree height of 27.5m at full scale). The mesh was bent twice on the uppermost third of the ring in order to decrease the local porosity and simulate the influence of the canopy. The aspect ratio and the arrangement of rings as shown on Figure 3, were identical for all ring categories. It is well known that the vertical wind profile, with its strong gradient located at the top of the canopy, is very similar to a mixing layer profile (Finnigan, 2000). The coherent large eddies generated in the shear region are responsible for strong wind gusts and sweeps, which play an important role in the momentum, temperature and mass exchange between the canopy and the atmosphere and in aerodynamic loads sustained by trees. Since the ultimate goal of the ECHO project is to study the vertical transport of emissions released from the canopy, properly reproducing the mean and turbulent velocity profiles through the canopy is of great importance. The strategy used here to find the appropriate arrangement of rings was to obtain by iterative preliminary testing (Aubrun and Leitl, 2004) the same aerodynamic properties of the flow, inside and above the canopy, as measured in the field.

The literature gathers different concepts to model forest or vegetation canopies from simple flexible sticks to architectural model trees (Ruck et al., 2012, Perret and Ruiz, 2013, Desmond et al. 2014). The conclusions showed that all these concepts are acceptable when one studies the flow right above the forest. It is more difficult to prove the ability to reproduce the flow within the canopy.

The model was placed in the large boundary layer wind tunnel of the Meteorological Institute of the University of Hamburg (Figure 4). Only neutral stability conditions can be replicated in this facility. The wind tunnel is characterised by a test section 18m long, 4m wide, and 2.5m high. The boundary layer was initialised by the presence of 10 turbulence spires at the entrance of the wind tunnel and of roughness elements covering the first 7.5 meters of the test section floor. The roughness elements were L-shaped metallic bluff obstacles of height = 50 mm, width of 30 mm and thickness of 2.5 mm, and were arranged in diamond arrays. This set-up determined the properties of the flow approaching the edge of the forest area.

The adjustable ceiling in the wind tunnel allowed for the compensation of the acceleration of the fluid due to the presence of the model in the test section and the growth of the boundary layer.

The horizontal static pressure distribution in the wind tunnel was measured with a pressure transducer (SETRA® 2671) through 11 flush-mounted pressure taps of 1 mm diameter, equally distributed along the test section at a height of 1.5m above ground. The pressure transducer was previously calibrated with a pressure balance. The guideline VDI 3783/12 (2000) recommends a tolerance threshold of 0.05 for the static pressure distribution criteria:

$$\left(\frac{\partial p}{\partial x}\delta\right)\bigg/\left(\frac{1}{2}\rho u_{\delta}^{2}\right)\le 0.05 \tag{1}$$

where is the longitudinal pressure gradient, δ is the boundary layer thickness, ρ the density of air and the flow velocity at the top edge of the modelled boundary layer. A maximum value of 0.015 was measured over the entire model.

Flow measurements were performed with a 2D fibre-optic Laser-Doppler-Anemometer (FVA-LDA, Dantec®) with 800 mm focal distance. The beam separation out of the beam expander was 75mm, leading to a measuring volume of $0.12\times0.12\times2.55mm^3$ ($0.036\times0.036\times0.765m^3$ in full scale). The flow was seeded with micro-particles of $2\mu m$ diameter. Zero- and first-order moments of the velocities were processed with at least 10 000 samples and a minimum averaging time of 120 seconds (10 hours in full scale with an identical reference velocity). In order to reproduce, in the wind tunnel, the tracer experiments performed in the field, a point source was set up in the model at the location of the source E2 from the field site (Figure 2). The source design ensured a release of gas in the horizontal plane with an impulse velocity lower than the local wind velocity. The height of the release was 13 ± 3 mm above ground (4±1m in full scale) for the source E2. The tracer gas used in the wind tunnel experiments was ethane. Measurements of instantaneous concentration were performed with a fast Flame-Ionisation-Detector (FID, Cambustion®). The sampling head consisted of a combustion chamber and a sampling needle (232 mm long, 0.3 mm inner diameter). The frequency resolution, for this case, was 60 Hz (0.2 Hz in full scale). The background concentration level of the flow was recorded with a slow FID and was removed from instantaneous measurements of the fast FID.

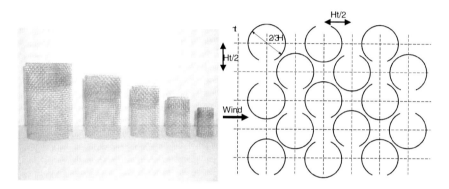

Figure 3. Open rings made from metallic mesh of (from left to right) 0.108, 0.092, 0.075, 0,058 and 0.042 m high and, arrangement of metallic rings representing the roughness caused by the trees in the forest area model. is the tree height.

Figure 4. The model of the ECHO site at a scale of 1:300 in the large boundary layer wind tunnel of the Meteorological Institute of University of Hamburg.

Both FID devices had a linear response in the range of the applied ethane concentrations. Nevertheless, they were calibrated before each measurement series with synthetic air and 3 certified calibration gases of different concentrations.

The flow rate of ethane released from the source E2 was controlled by a Brooks® mass flow controller (5851 S-series). The mass flow controller was calibrated with a specified volume meter Brooks Vol-U-meter ®.

FIELD VERSUS WIND TUNNEL RESULTS

No field data of the approach flow of the Juelich forest are available. Nevertheless, the area surrounding the forest stand comprises farmlands and grasslands, which belong to moderately rough surfaces. Based on a compilation of field data, the German guideline VDI 3783/12 (2000) provides requirements to properly replicate the neutral atmospheric boundary layer up to a height of 100m. According to this guideline, for a moderately rough surface, the power law exponent should be between $0.12 < \alpha < 0.18$, the roughness length between $5\text{mm} < z_0 < 100\text{mm}$ and the displacement height $d_0 \approx 0$. The parameters α and then determine the range of the turbulence intensity profiles.

The modelled boundary layer was measured at 0.5m (150m in full scale) upstream of the edge of the forest area. The best fit to the measured wind profile with an exponential function is obtained with a power law exponent α of 0.19. The wind profile fits to a logarithmic function within the lowest part of the profile. The extrapolation of the logarithmic law towards the zero-value of velocity gives the associated roughness length, $z_0 = 0.2$m. The power law exponent and the roughness length are slightly higher than the recommended ranges but are consistent with each other. The friction velocity is estimated to $u_* = 0.33 \text{ m s}^{-1}$ by averaging the shear stresses measured in the surface layer. The turbulence intensity profiles are enclosed within the ranges advised by VDI 3783/12 (2000) for a height up to 90m.

The spectra of turbulent fluctuations measured in the wind tunnel corresponding to 25 and 50m full scale above ground are validated against the empirical laws based on field measurements over a flat uniform terrain under neutral stability conditions and published in Kaimal and Finnigan (1994). The vertical fluctuations in the wind tunnel are slightly larger than those expected from the empirical laws.

The longitudinal integral length scale was calculated applying the autocorrelation method on the time series of velocity fluctuations at different heights above ground. the vertical distribution of is validated against the empirical law proposed by Counihan (1975):

$$L_u(z) = C(z_0) \, z^{1/n(z_0)} \tag{5}$$

with and for a roughness length of $z_0 = 0.2$ m. The agreement is satisfying.

Flow above and Within the Forest Canopy

As the flow meets the forest area, the atmospheric boundary layer redevelops under very rough conditions due to the presence of the trees. The inhomogeneity of the forest distribution leads to a very complex configuration. The structure of the turbulent flow in the presence of the forest canopy was studied at two locations of the forest-site, where meteorological field data were available: at the meteorological tower and the 17m-mast.

At the meteorological tower, only data characterised by a wind direction WD between 268°<WD<272°, a mean velocity higher than 4m s^{-1} at 30m above ground and neutral stability conditions (L =Monin-Obukhov length) were selected. From this remaining 20 hours of data, mean velocity and turbulence profiles, turbulent spectra and integral length scales were computed.

The selected field data were compared with the equivalent wind tunnel experiments (Figure 5 and 6). In these figures, the height above ground is non-dimensioned by the local tree height and the flow measurements obtained at (the highest measurement location in field) are used as reference values.

The wind profiles (Figure 5a) measured in the wind tunnel at the 17m-mast is typical for a dense forest canopy since the velocity is very low within the canopy and presents a very strong gradient in the upper part of the canopy (Finnigan, 2000). The field data show similar properties but the mean velocity measured inside the canopy was even smaller. Since the meteorological tower is located in a clearing, the lower part of the boundary layer has a local re-development, which is slightly underestimated in the wind tunnel.

The vertical distributions of the standard deviations of the velocity (Figure 6b) show that the turbulence properties of the flow are completely different in the two distinct regions, inside and above the canopy: In both regions, the standard deviation is nearly constant and the transition between these two states takes place between $0.85 < z/Ht < 1.1$. The first region is dominated by the local effect of the forest whereas the second is representative of the development of the atmospheric boundary layer on a very rough surface. For the reasons explained previously, this feature is not as distinct in the case of the meteorological tower; the transition region ranges is between $0.5 < z/Ht < 1.5$. A so strong vertical gradient in the standard deviation profile, as measured at the 17m-mast, is only obtained for dense to very dense canopies (Finnigan, 2000).

The non-dimensional standard deviations of longitudinal velocity measured in the wind tunnel and in the field are in very good agreement. The non-dimensional standard deviations of lateral and vertical velocities measured at the meteorological tower are slightly overestimated in the model. On the other hand, the stress ratios and measured right above the canopy in the wind tunnel stand within the scatter of values found in literature (Finnigan, 2000).

The Dispersion Process

The comparison of tracer-gas experiments performed in field and in the wind tunnel with the source E2 is presented on the Figure 7. The dimensionless concentration is plotted, where

$$c^* = \frac{\overline{c}\, U_{ref}\, L_{ref}^2}{Q} \tag{6}$$

In the wind tunnel experiments, [-] is the time-averaged volume ratio of tracer, [m s^{-1}] is the velocity measured at the location of the meteorological tower at = 0.1m (i.e., 30m at full scale) above ground and [m³ s^{-1}] is the volume flow of the point source. In the field experiments, [g m^{-3}] is the time-averaged mass concentration of tracer gas, [m s^{-1}] is the velocity measured at the meteorological tower at = 30m above ground and [g s^{-1}] is the mass flow of the point source.

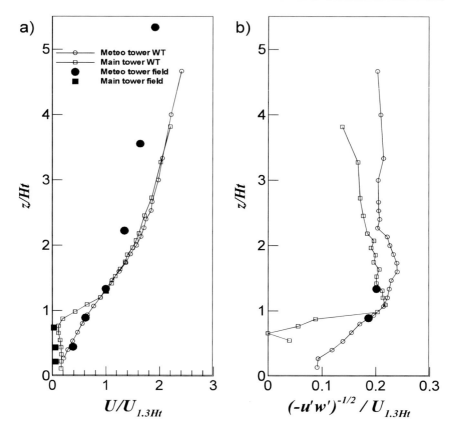

Figure 5. Vertical profiles measured at the meteorological tower and at the 17m-mast in field and in the wind tunnel. a) time-mean velocity non-dimensioned with the time-mean velocity at 1.3 Ht above ground $U_{1.3Ht}$, and b) shear stresses $-\overline{u'w'}$ non-dimensioned with the time-mean velocity at 1.3 Ht above ground $U_{1.3Ht}$.

The wind direction in the wind tunnel is fixed to 270° whereas during the field campaigns it showed a scatter of up to 30°. This fact could justify the obvious difference of the plume axis between the field and the wind tunnel data. Nevertheless, the dimensionless concentration c^* measured at a certain distance of the source has the same level of magnitude in the field and in the wind tunnel. Considering the differences in the meteorological conditions between the field and the wind tunnel experiments, and knowing that the dispersion process is very sensitive to the terrain roughness (Roberts et al., 1994), obtaining the same order of magnitude in the field and in the wind tunnel is satisfying. Nevertheless, additional tests are needed before to conclude that the physical model of the forest area can reproduce the dispersion process inside the forest canopy.

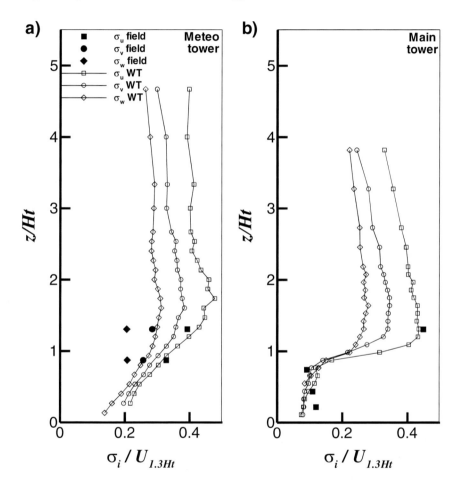

Figure 6. The standard deviations of the longitudinal, lateral and vertical velocities non-dimensioned with the time-mean velocity at $1.3\,Ht$ above ground $U_{1.3Ht}$, measured in field and in the wind tunnel a) at the meteorological tower and b) at the 17m-mast.

FOOTPRINT EXPERIMENTS IN WIND TUNNEL

A point source of ethane was travelled with the traverse system all over the model area. The point source was designed to release the tracer-gas horizontally with an exhaust velocity similar to the minimal velocity measured inside the modelled forest area. The fractions of ethane that reached the 4 sampling points of the main tower (located at $z/Ht_{main} = 0.3$), ($z/Ht_{main} = 0.7$), ($z/Ht_{main} = 1$) and ($z/Ht_{main} = 1.3$) were measured with slow Flow Ionisation Detectors.

Two purposes had motivated the footprint experiments. The first one was to determine the origin of air masses, which were sampled at the main tower. In that case, the point source was travelled within 3 different vertical layers located at $x/Ht_{main} = -11, -46$ and -90 upstream of the main tower respectively. The 2 first layers cross laterally the forest area and the last one is located right upstream of the forest area. The second aim (not presented in this paper) was to estimate the proportion of the emissions released by the forest that reaches the sampling points of the main tower. The point source was travelled all over the forest area at a height of 80% of the height of the local trees. This release height had been chosen because the maximum in the vertical profile of biogenic emission strength obtained in a forest area is located at 80% of the tree height.

The figure 8 presents the distribution of the probability of the origin, in a vertical layer located upstream of the forest area, of air masses sampled at the main tower at and respectively. The coordinates of the main tower are $x_{main} = y_{main} = 0$. In the case of a homogeneous and infinite forest, the distribution of the origin of air masses would have been symmetric in the Y-axis. In both layers, the distributions are shifted laterally under the effect of the large-scale flow distortion due to the presence of the finite forest area and of the local flow distortions due to its inhomogeneities. The air masses come predominantly from the ground area and are deflected up when they encounter the forest area. The very high level of turbulence present at the top of the canopy causes a strong vertical and lateral dispersion of air masses. The distribution of the provenience of air masses sampled inside the canopy is wider than above the canopy because the flow is more subjected to the local geometry of the forest.

Physical Modelling of Biogenic VOCs Emission and Dispersion ... 17

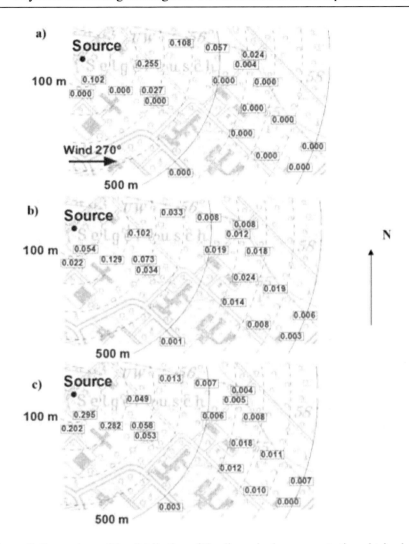

Figure 7. Comparison of the distribution of the dimensionless concentration obtained in the wind tunnel and in field during the tracer-gas experiments. a) in the wind tunnel (averaging time of 25 hours full scale), b) in field on the 13th September 2000 (80-min average) and c) in field on the 12th June 2001 (60-min average).

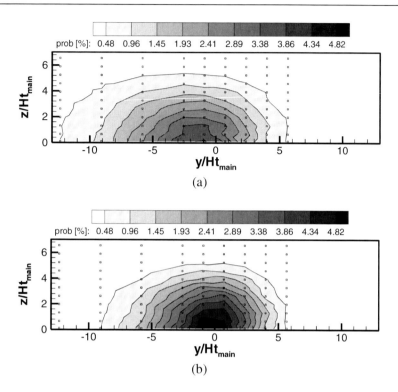

Figure 8. Probability distribution of the origin in the vertical layer $x/Ht_{main} = -90$ of the air masses sampled at the main tower at a) P_l, b) P_h.

The figure 9 presents the distribution of the probability of the origin, in a vertical layer located at $x/Ht_{main} = -46$ upstream of the main tower, of the air masses sampled at the main tower at and respectively. This layer is located inside the forest area and the distributions show that the maximum of provenience is located at $z/Ht_{main} = 1$. The air masses are not subjected anymore to an upward deflection and are travelling horizontally.

The figure 10 presents the distribution of the probability of the origin, in a vertical layer located at $x/Ht_{main} = -11$ upstream of the main tower, of the air masses sampled at the main tower at P_l, P_{ml}, and respectively. The distributions present two different poles of provenience of air masses. That means that the flow follows two different paths, under the effect of the inhomogeneities of the forest. One pole is located at $z/Ht_{main} = 1$ and close to the Y-axis and the second one at the ground level and strongly laterally shifted. The first pole still corresponds to the horizontal convection of air masses. The second one is generated by the proximity of the clearing surrounding the meteorological tower. The boundary layer locally redevelops, generating a flow acceleration inside the clearing, which results in a local dynamical impulse inside the canopy. At the same time, the end of the clearing causes a blockage that forces an upward deflection of air masses. Consequently, the air masses travel at higher altitudes in the canopy. The combined phenomena justify that the air masses, which were close to the ground at $x/Ht_{main} = -11$, reach all the sampling stations at the main tower.

The air masses sampled at the two measurement stations located inside the canopy (at and P_{ml}) are predominantly coming from the ground whereas the ones sampled at the top of the canopy (P_{mh} and P_h) are evenly dominated by both poles of origin.

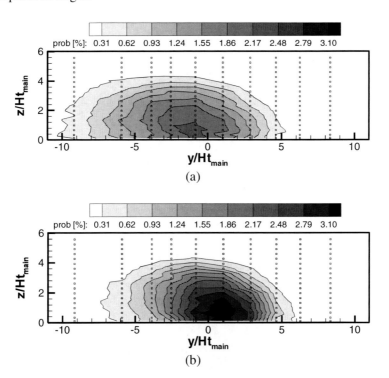

Figure 9. Probability distribution of the origin in the vertical layer $x/Ht_{main} = -46$ of the air masses sampled at the main tower at a) P_l, b) P_h.

The travel time of the air masses through the forest area had been assessed. To do so, the delay between the start of release of ethane from the point source and the start of detection of non-zero concentration at the sampling points was measured. This operation had been repeated 50 times for each configuration in order to build a reliable statistical result. The travel time was non-dimensioned by the travel time due only to advection τ_{adv}. Figure 10 shows the vertical profile of for the most likely sampled air masses from each of the three vertical layers sampled at and P_h. The height where corresponds to the averaging travel altitude. The figures 10a and 10b show that the average travel altitude deduced from the long travel distance ($x/Ht_{main} = -90$ and -46) is the tree height. These results lead to the conclusion that the edge of the forest produces a sudden upward deflection of the air masses, which travel horizontally after a short transient distance. The fact that the average travel altitude is independent of the sampling height signifies that the vertical dispersion of the air masses due to the vertical turbulent fluctuations is very strong.

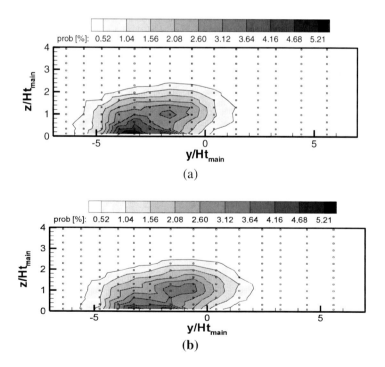

Figure 10. (Continued).

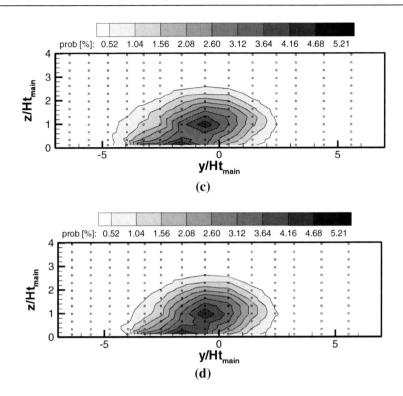

Figure 10. Probability distribution of the origin in the vertical layer $x/Ht_{main} = -11$ of the air masses sampled at the main tower at a) P_l, b) P_{ml}, c) P_{mh}, d) P_h.

The average travel altitudes derived from shorter distances, where air masses are travelling through a more inhomogeneous region ($x/Ht_{main} = -11$), illustrate the complexity of the local flow. The air masses coming from the top of the canopy ($z/Ht_{main} = 1$) continue to travel horizontally ($0.9 < z/Ht_{main} < 0.95$) and are vertically dispersed under the effect of the vertical turbulent fluctuations. The air masses coming from the ground area and sampled above the canopy have an average travel altitude of $z/Ht_{main} = 0.9$. That confirms the previously issued explanation about the deflection caused by the clearing surrounding the meteorological tower. On the other hand, the air masses coming from the ground area and sampled at shows a larger range of travel altitudes ($0.6 < z/Ht_{main} < 0.85$) located inside the canopy. Nevertheless, the ratio is close to 1 all over the canopy, which also leads to the conclusion that the travel altitude is from the ground up to 0.85.

The travel time of the air masses coming from the ground at -11 is 1.5 larger than the travel time of air masses coming from the top of the canopy. Thus, the probability that the VOC chemically react with each other during the travel is higher from air masses coming from the ground area.

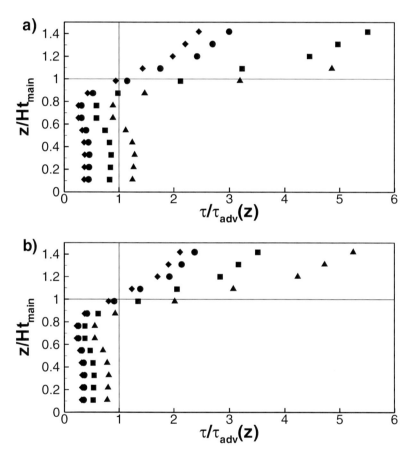

Figure 9. Travel times of the air masses non-dimensioned by the advection time profiles for air masses sampled at a) P_l, b) P_h. ♦ $x/Ht_{main} = -90$, $y/Ht_{main} = -0.8$, 0.3. ● $x/Ht_{main} = -46, 1, 1$. ■ $x/Ht_{main} = -11$, $y/Ht_{main} = -0.6, 1$. ▲ $x/Ht_{main} = -11$, (a) -3.2 (b) -1.6, $z/Ht_{main} = 0.1$.

CONCLUSION

Within the joint research project ECHO a wind tunnel model of the complex field site has been set up. In order to check whether the physical model was able to replicate the flow properties of the atmospheric boundary layer in the presence of a complex forest canopy, mean and turbulent velocity profiles obtained in the field and in the wind tunnel at two different locations were successfully compared. This local agreement must be complemented by a more global comparative approach. Consequently, the comparison of the dispersion process inside the forest area was performed by carrying out similar tracer experiments in the field and in the wind tunnel.

Due to the inevitable variations of the boundary conditions during the field experiments, the results of the field and of the wind tunnel experiments could not be compared point by point at the different sampling positions. Nevertheless, the dimensionless concentrations were of the same order of magnitude. Since the dispersion process is very sensitive to the terrain roughness, obtaining the same order of magnitude is a satisfying result. The model were then used to perform extensive and high-resolution experiments that are not feasible in the field, in order to investigate the consequence of the interactions between the canopy and the atmospheric boundary layer on the biogenic emissions transport. Footprint experiments were performed to provide information on the transport of the air masses through the forest area and on the origin of the biogenic emissions sampled at the measurement tower during the field campaigns. The travel time of these emissions were also assessed. Both information gave some hints to the chemists to determine whether and how, the VOC can be chemically processed during their travel within the canopy.

ACKNOWLEDGMENTS

This project was funded by the German Ministry for Education and Research within the Atmospheric Research Program AFO2000 under grant No. 07ATF47.

REFERENCES

Aubrun, S; Leitl, B; Schatzmann, M. Physical modelling of a finite forest area – Transport of biogenic emissions. International Conference 'Wind Effects on Trees' University of Karlsruhe, Germany, September 16-18, 2003

Aubrun, S; Koppmann, R; Leitl, B; Möllmann-Coers, M; Schaub, A. Physical modelling of a complex forest area in a wind tunnel – Comparison with field data. *Agricultural and Forest Meteorology*, 129, 121-135, 2005

Aubrun, S; Leitl, B. Development of an improved physical modelling of a forest area in a wind tunnel. *Atmospheric Environment*, 38/18, 2797-2801, 2004.

Beger, G. Windkanaluntersuchungen an modellierten Waldkomplexen. Wiss. Z. Techn. Univers. *Dresden*, 32, H. 5, 1983.

Brunet, Y; Irvine, MR. The control of coherent eddies in vegetation canopies: streamwise structure spacing, canopy shear scale and atmospheric stability. *Boundary Layer Meteorology*, 94, 139-163, 2000.

Counihan, J. Adiabatic atmospheric boundary layers: a review and analysis of data from the period 1880-1972. *Atmospheric Environment*, 9, 871-905, 1975.

Desmond, CJ; Watson, SJ; Aubrun, S; Ávila, S; Hancock, PE; Sayer, A. A study on the inclusion of forest canopy morphology data in numerical simulations for the purpose of wind resource assessment. *J. Wind Eng. Ind. Aerodyn*, 126, 24–37, 2014

Engineering Sciences Data Unit Characteristics of atmospheric turbulence near the ground. Part II: Single point data for strong winds (neutral atmosphere). Item n° 85020. ESDU International, London 1985.

Finnigan, JJ. Turbulence in plant canopies. *Annual Review of Fluid Mechanics*, 32, 519-571, 2000

Finnigan, JJ; Shaw, RH. A wind tunnel study of airflow in waving wheat: an Empirical Orthogonal Function analysis of the large-eddy motion. *Boundary Layer Meteorology*, 96, 211-255, 2000

Gardiner, BA. Wind and wind forces in a plantation spruce forest. *Boundary-Layer Meteorology*, 67, 161-186, 1994

Hall, DJ; Walker, S; Spanton, AM. Dispersion from courtyards and other enclosed spaces. *Atmospheric Environment*, 33, 1187-1203, 1999

Irvine, MR; Gardiner, BA; Hill, MK. The evolution of turbulence across a forest edge. *Boundary Layer Meteorology*, 84(3), 467-496, 1997

Kaimal, JC; Finnigan, J. *Atmospheric Boundary Layer Flows*. Oxford University Press. 1994

Kaimal, JC; Wyngaard, JC; Izumi, Y; Coté, OR. Spectral characteristics of surface-layer turbulence. *Quaterly Journal Royal Meteorology Society*, 98, pp. 563-589, 1972

Katul, GG; Albertson, JD. An investigation of high-order closure models for a forested canopy. *Boundary Layer Meteorology*, 89, 47-74, 1998

Marshall, BJ; Wood, CJ; Gardiner, BA; Belcher, RE. Conditional sampling of forest canopy gusts. *Boundary Layer Meteorology*, 102, 225-251, 2002

Massman, WJ; Weil, JC. An analytical one-dimensional second-order closure model of turbulence statistics and the Lagrangian time scale within and above plant canopies of arbitrary structure. *Boundary Layer Meteorology*, 91, 81-107, 1999

Möllmann-Coers, M; Klemp, D; Mannschreck, K; Slemr, F. Determination of anthropogenic emissions in the Augsburg area by the source-tracer-ratio method. *Atmospheric environment*, 36 Supplement No. 1, 95-107, 2002

Morse, AP; Gardiner, BA; Marshall, BJ. Mechanisms controlling turbulence development across a forest edge. *Boundary Layer Meteorology*, 103, 227-251, 2002

Novak, MD; Warland, JS; Orchansky, AL; Ketler, R; Green, S. Wind tunnel and field measurements of turbulent flow in forests. Part 1: Uniformly thinned stands. *Boundary Layer Meteorology*, 95, 457-495, 2000

Pasquill, F. *Atmospheric Diffusion*. John Wiley & Sons, New York, 1974

Perret, L; Ruiz, N. SPIV analysis of coherent structures in a vegetation canopy model flow. *Coherent Flow Structures at Earth's Surface*, First Edition. John Wiley & Sons, New York, 2013.

Plate, EJ. Studies in Wind Engineering and Industrial *Aerodynamics*, Vol.1. Engineering Meteorology. Elsevier Scientific Publishing Company, p.740, 1982

Raupach, MR; Finnigan, JJ; Brunet, Y. Coherent eddies and turbulence in vegetation canopies: the mixing-layer analogy. *Boundary-Layer Meteorology*, 78, 351-382, 1996

Rannik, Ü; Markkanen, T; Raittila, J; Hari, P; Vesala, T. Turbulence statistics inside and over forest: influence on footprint prediction. *Boundary Layer Meteorology*, 109, 163-189, 2003

Roberts, PT; Fryer-Taylor, REJ; Hall, DJ. Wind-tunnel studies of roughness effects in gas dispersion. *Atmospheric Environment*, 28/11, 1861-1870, 1994

Ruck, B; Frank, C; Tischmacher, M. On the influence of windward edge structure and stand density on the flow characteristics at forest edges. *Eur J Forest Res,* 131, 177–189, 2012

Snyder, WH. Guideline for fluid modelling of atmospheric diffusion. US Environmental Protection Agency. EPA-600/8-81-009, p. 185, 1981

Teuchert, D. Ein bodennahes Strömungsmodell für den Waldbestand – Entwicklung und Anpassung eines diagnostischen Strömungsmodells für den operationellen Betrieb. PhD. thesis from the University of Cologne. *Report from Forschungszentrum Jülich,* volume 3964, 2002

VDI-guideline 3782/1, Gaussian plume model for air quality management. Beuth Verlag, Berlin, 2001

VDI-guideline 3783/12 Physical modelling of flow and dispersion processes in the atmospheric boundary layer – Application of wind tunnels. Beuth Verlag, Berlin, 2000

Villana, MG; Schmid, HP; Su, HB; Hutton, JL; Vogel, CS. Turbulence statistics measurements in a northern hardwood forest. *Boundary Layer Meteorology,* 108, 343-364, 2003.

In: Volatile Organic Compounds
Editor: Khaled Chetehouna

ISBN: 978-1-63117-862-7
© 2014 Nova Science Publishers, Inc.

Chapter 2

ESTIMATION OF VOCs EMISSIONS DURING THE WILDLAND FIRES FROM 1995 TO 2009 IN CORSICA

T. Barboni[*], P. A. Santoni and F. Bosseur

SPE– UMR 6134 CNRS, University of Corsica, Campus Grimaldi, BP 52, 20250 Corte, France

ABSTRACT

The countries of the Mediterranean basin are particularly affected by fires, which travel several thousand hectares per year. France is particularly vulnerable to wildland fires and mainly Corsica, where 84,000 hectares of maquis and forest have been burnt from 1995 to 2009. This paper present an estimation of volatile organic compounds (VOC) emitted in Corsica during this period. We first deal with the identification and the quantification of the main VOCs found in wildfire smoke. These results were obtained on a combustion chamber of 0.4 m^3 containing fuel submitted to an epiradiator and a sampling pump with an adsorbent tube (Tenax TA). The analysis is carried out with an Automated Thermal Desorption (ATD) coupled with a Gas Chromatography (GC) and with two detectors: Mass Spectrometry (MS) and Flame Ionization Detector (FID). 71 VOCs were identified. These emissions represent 25.4 kg/ha burnt. An extrapolation of these results for fires in Corsica leads to almost 1.92 Gigagrams of VOCs emitted into the atmosphere from 1995 to 2009.

---

[*] Corresponding author: Email: barboni@univ-corse.fr.

Forest fires increase the potential risk for the population long time exposure.

Keywords: Corsica Wildfires - Emission factors – VOCs

1. INTRODUCTION

Every year, millions hectares of forest are burnt by wildland fires worldwide. The countries of the Mediterranean basin are particularly affected by these fires, which travel several thousand hectares per year and cause severe ecological, economic and social damages. France with its 15 million hectares of forest is particularly vulnerable to wildland fires and mainly in Corsica, where the Corsican maquis is exceptionally extensive over 200 000 hectares or 20% of the surface of Corsica. The forest fires may affect local, regional and global environments. Wildfires cause damage to ecosystems, significant economic losses and human tragedies. The consequences of these fires are many: climate, soil erosion and loss of biodiver. During the combustion, many pollutants are formed. Due to the effects of these pollutants on health, it is important to quantify them in smoke to assess the potential risk they cause to staff and the populations exposed. The composition of smoke is variable depending on the fuel, load, its geometry, humidity and weather conditions.

Wildfires produce gases and particles that affect the composition of the atmosphere. Fires affect carbon sequestration by forests, and the greenhouse gases emitted from forest fires directly affect the global and regional carbon cycles (Narayan et al. 2007). Conversely, climate change will cause an increase in drought situations and thus increased risk of fire as well as severe alterations in fire regimes (Flannigan et al.1998, Stocks et al. 1998, 2003). Smoke is composed primarily of carbon dioxide, water vapor, carbon monoxide, particulate matter, hydrocarbons and other organic chemicals, nitrogen oxides, trace minerals and several thousand other compounds (Barboni et al. 2010b, Evtyugina et al. 2013, Gouw et al. 2006, Miranda 2004, Statheropoulos and Karma 2007, Urbanski 2008, Yokelson 2007, Ward 1998).

Volatile organic compounds (VOCs) include hydrocarbons, aromatic hydrocarbons, oxygenated compounds such as alcohols, aldehydes, ketones, furans, carboxylic acids, esters, and also isoprenoid compounds (Andreae and Merlet 2001, Barboni and Chiaramonti 2010, Barboni et al. 2010 a,b, 2011, Chetehouna 2009, Dost 1991, Evtyugina 2014, Friedli et al. 2001, Gouw et al.

2006, Reisen and Brown 2009, Sinha 2003, 2004, Urbanski 2008, Yokelson 2007, 2008). Characterization of primary VOC emissions is needed to accurately initialize photochemical models for wildfire impact assessment from local to global scales. There are very few studies on the characterization of these VOCs emitted during burning of vegetation and less concerning the European vegetation (Barboni et al. 2010b, Evtyugina et al. 2013, 2014, Miranda et al. 2010, Statheropoulos and Karma 2007). Among the gases and particles emitted, VOCs represent an important part of potential risk to human health, because these compounds can penetrate the higher respiratory tracts down to the lungs.

This chapter presents many results that improve the knowledge on the identification and the quantification of VOCs in smoke from wildfire. To this end, an experimental device was used that is based on a combustion chamber of 0.4 m3 containing fuel submitted to an epiradiator and a sampling pump with an adsorbent tube (Tenax TA). The analysis is carried out with an Automated Thermal Desorption (ATD) coupled with a Gas Chromatography (GC) and with two detectors: Mass Spectrometry (MS) and Flame Ionization Detector (FID). An extrapolation of these results for fires that occurred in Corsica from 1995 to 2009 allows us to estimate the amount of VOCs emitted into the atmosphere during this period. These emissions due to wildfires lead to increase the potential risk for the population long time exposure.

2. METHODS

2.1. Fire Data

The fire data used in this paper have been obtained from the national France database Promethee (http://www.promethee.com/). Since 1973, this database records the fire history (ignition point location, surface burnt, day, i.e.) for the 15 departments of the south-east of France. All informations are collected by different services (firefighters, foresters, and police) for fires for which control actions were done.

2.2. Calculation of Emissions

Emission was estimated following the standard method for estimating emissions from burnt biomass used by many authors (French et al. 2004,

Kasischke and Bruhwiler 2003, Lü 2006, Seiler and Crutzen 1980). According to this method, the total fuel consumed during fires (FCT), is given by:

$$FCT = FLS \times CS \tag{1}$$

where FLS is the surface fuel in the area burnt, and CS is the respective combustion completeness (the ratio of fuel consumption to total available fuels) of surface fuels. CS was approximate to 0.9. This value was obtained in laboratory for many studies done with an oxygen consumption calorimeter (Barboni et al. 2012, Proterina-C 2011, Santoni et al. 2011).

The amount of a specific trace gas released during fires (ES) can be calculated as:

$$ES = FCT \times EF \tag{2}$$

where EF is the emission factor for the gas species. EF is usually defined in terms of grams of trace gas emitted per kilogram of dry matter consumed during fires (FCT). The total amont of VOCs (TVOC) emitted by the fire is:

$$\boldsymbol{TVOC} = \sum_{i=1}^{n} ES_i \tag{3}$$

Where i represent a compound and n is the total number of compounds considered.

2.3. Identification and Determination of EF

The experimental device is based on a combustion chamber of 0.4 m^3 containing fuel submitted to an epiradiator and a sampling pump with an adsorbent tube (Tenax TA). Five species representative of the mediterrannean maquis of Corsica were burnt: needles of *Pinus nigra* ssp *laricio* (Poir.) Maire var. *corsicana* (Loud.), needles of *Pinus pinaster* (Ait), leaves of *Cistus monspelliensis* (L.), leaves of *Arbutus unedo* (L.) and leaves of *Erica arborea* (L.).

The analyses were carried out immediately at the laboratory using an Automatic Thermal Desorber Perkin Elmer® ATD turbomatrix. A sorbant was brought to 280°C during 10 minutes and a carrier gas flushed the sample towards a cold trap at -30°C.

In a second step, the cold trap (22 cm, 0.53 mm i.d. (Supelco)) was programmed for an increase in temperature by -30 to 280°C at 40°C.s^{-1} then held an isotherm at 280°C during 3 minutes, desorbing the compounds to the chromatograph. The injector temperature: 280°C, splitless, energy ionization: 70 eV, electron ionization mass spectra were acquired over the mass range 35-350 Da.

The chromatograph and the mass spectrometer are Perkin Elmer® Clarus 500 apparatus. The carrier gas was helium which exerted a pressure of 20 psi at the head of the column. The chromatograph was equipped with dual flame ionization detector (FID) system, the mass spectrometer and two fused-silica capillary columns (Rtx-1, dimethylsiloxan, length: 60 meters and internal diameter: 0.22 μm). FID detector temperatures were maintained at 280 °C. Samples were injected in the splitless mode, using helium as carrier gas (1 mL/min).

Detection was done by a quadripolar analyzer made up of an assembly of four parallel electrodes of cylindrical section. The temperature program of the analysis orders an increase in temperature from 50 to 260°C at 2°C per minute with a stage at 260°C during 10 minutes. Transfert line temperature was at 280°C.

The methodology carried out for identification of individual components was based:

(a) on comparison of calculated retention indices (I) of compounds were determined relative to the retention times of series of n-alkanes (C_5-C_{30}) with linear interpolation (using Van den Dool and Kratz equation and software from Perkin-Elmer), apolar (Ia) columns, and those of authentic compounds or literature data (König et al. 2001, NIST 2005);

(b) on computer matching with commercial mass spectral libraries (NIST 1999) and comparison of mass spectra with those of our own library of authentic compounds or literature data (Adams 2001, König et al. 2001). The benzene (Commercial source: Restek) internal standard was used for the quantification.

3. RESULTS

3.1. General Analysis of the Forest Fires in Corsica over the Period 1995-2009

More than 2,500 plant species grow naturally in Corsica and among these plants more than 160 are specific to Mediterranean islands and 135 are endemic to Corsica. The main vegetation of Corsica is the maquis but we must also consider the forests, wetlands and grasslands. Corsican maquis is exceptionally extensive over 200 000 hectares or 20% of the surface of Corsica. The wildfires are most important during the 3 summer months (July-September).

During the period 1995-2009, the total number of fires was about 12,400 that burnt more than 84,000 ha (Figure 1). The average number of fire was 826 and the average annual area burnt was 5608 ha, ranging from 370 ha in 2008 to 27355 ha in 2003 (Figure 1). We observe with a trend curve that the annual total number of fires decreases during the period 1995-2009. The annual area burnt varied widely from year to year. The biggest peaks of burned area were observed in 2000 and 2003, and these years were the severe drought and the incendiary pressure was the highest during this period. We observe there is a link between the area burnt and the driest seasons. However, since 2003, the trend is downward with the work of prevention, management of forest areas and less critical conditions.

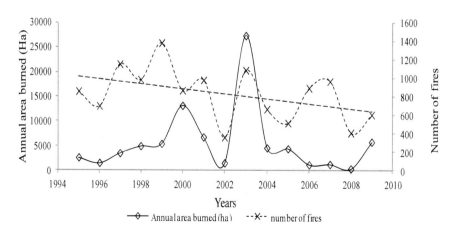

Figure 1. Annual area burnt and number of fires in Corsica during the period 1995-2009.

Table 1. Analysis of fires in Corsica according to area burned

	0 to 50 ha	50 to 100 ha	> 100 ha
Number of fires	12231	54	101
% of total fires	98.75	0.44	0.82
Area burnt (ha)	15659	3848	64608
% of total area burnt	18.62	4.57	76.81
Average area burnt per fire (ha)	1.2	71.3	639.7

An analysis of fires in Corsica may be realized according to area burned. To do this, we were cut the fires into 3 categories according to the fires area burned. These categories are: 0 to 50 ha: small fires to medium fires, 50 to 100 ha: medium fires to big fires, 100 ha and more: big fires. Table 1 provides the number of fires and surface burnt according to this classification. A point that deserve mention is that the majority of fires are controlled and do not exceed 50 hectares of surface burnt (Table 1). We can see that a very small number of fires representing 0.82% of the total number of fires are responsible of most of the burnt areas (76.81% of the surface burnt). The big fires (> 100 ha) are the fires that cause the more damage and it is essential identify the causes of these large fires for to reduce their damage.

A large fire is initially an ordinary fire like others that for special reasons like extreme meteorological conditions or very few available crews to do suppression attacks could not be contained. Figure 2 displays as example the maps of the ignition points and surface burnt by the big fires in Corsica from 1995 to 2009 (Figure 2).

3.2. Identification, Emission Factor and Estimation of Total VOCs Emitted during the Period 1995-2009 in Corsica

71 COVs were identified by ATD-GC/MS and ATD-GC-FID (Barboni et al. 2010b). These are compounds in C_5 to C_{30}. EF represent in average values of 1357 ± 265 mg.Kg$^{-1}_{dw}$ (dw: dry weight) for the five vegetation species examined (Table 2). The process identified 30 non-terpenic hydrocarbons, 28 non-terpenic oxygenated and 13 terpenic compounds. Hydrocarbon compounds were 602.4 ± 78.5 mg.Kg$^{-1}_{dw}$ including benzene and benzene derivatives, alkanes, alkenes. Other ring systems included furans, 1H-indene, pyrroles, and naphthalene and theirs derivatives. Oxygenated compounds were 665.4 ± 106.0 mg.Kg$^{-1}_{dw}$ phenol and phenol derivatives, organic acids, esters

and aldehyde compounds. Also, 13 terpenes were 89.6±120.6 mg.Kg$^{-1}_{dw}$ including five monterpenes compounds and 8 sesquiterpenes compounds. Table 1, we distinguish the total volatile organic compounds (TVOC) and biogenic volatile organic compounds (TBCOV). The TBCOV represent the compounds emitted naturally by plants and do not cause real effects on human health but are ozone precursors.

This chapter presents an estimation of VOC produced by the fires during 1995 to 2009 in Corsica. Both parameters can cause many errors, the combustion efficiency to a value of 0.9 (Barboni et al. 2012, Santoni et al. 2011) and the fuel load of a value of 2 kg.m^{-2} for maquis (Proterina-C, 2011).

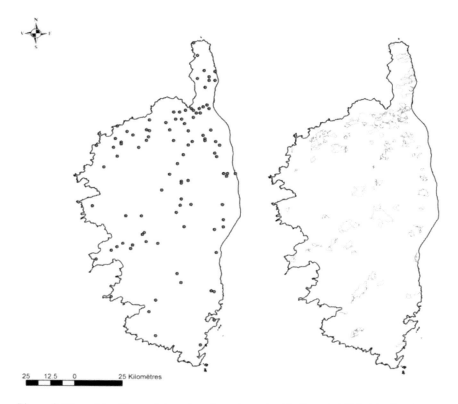

Figure 2. Map of ignition points and surfaces burnt by big fires (>100 ha) in Corsica over the period 1995-2009.

Table 2. Mean values and standard deviations of VOCs emissions from leaves or needles of Pinus larico (PL) *Pinus pinaster* (PP), *Cistus mospeliensis* (CM), *Arbutus unedo* (AU) *and Erica arborea* (EA)

(mg.Kg^{-1}$_{dw}$)	EF$_{P.L}$	EF$_{P.P}$	EF$_{C.M}$	EF$_{A.U}$	EF$_{E.A}$	EF$_{average}$
Terpenic compounds	277.9	142.8	27.5	0	0	89.6±120.6
Hydrocarbons compounds	914	511.8	518.5	587.9	479.9	602.4±178.5
Oxygenated compounds	603.1	541.5	744.2	802	636.2	665.4±106.0
TVOC + TBVOC (g.Kg^{-1}$_{dw}$)	1.795	1.196	1.290	1.390	1.116	1.358±0.265
TVOC (g.Kg^{-1}$_{dw}$)	1.517	1.053	1.263	1.390	1.116	1.268±0.191

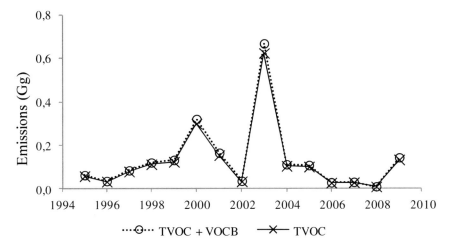

Figure 3. TVOC and TVOC+VOCB emissions from forest fires in Corsica during the period 1995 - 2009.

The mean total volatile organic compounds (TVOC, calculated with equations 1 to 3) released per year by forest fires in Corsica is about 0.128 Gg, ranging from 0.008 Gg in 2006 to 0.624 Gg in 2003 at highest. With the BVOC, the sum represents 0.137 Gg for the mean annual emission. These emissions from forest fires varied significantly from year to year due to annual variations in the area burnt (Figure 3). The years 2000 and 2003, corresponds to the highest emissions of TVOC. For these years, the emissions were respectively 35 and 74 times greater than in 2006. There is a link between the emission of VOCs and the driest seasons. In 2000 and 2003 there were severe droughts and as a consequence big fires occurred that for which maximum VOCs were emitted. The total TVOC emitted during the period 1995-2009 was 1.92 Gg. The big fires (> 100 ha) represent 1.475 Gg of TVOC which corresponds to 76.8% of the total amount emitted during this period. The big fires are therefore the largest emitters of VOCs. Particular attention must be paid to the reason of the occurrence of big fires. For instance extreme meteorological (high ambient temperature and low air-humidity) conditions were identified as a criticalities for big fires. Hence, in order to decrease the TVOC, one must obviously reduce the number of big fires and increase the prevention during days with extreme meteorological conditions.

CONCLUSION

In Corsica, 12,400 fires occurred during the period 1995-2009. They burnt more than 84,000 ha. The average annual area burnt was 5608 ha and the number of fires was 826. We estimated the direct emissions of VOCs from Corsican Fires during this period. Our results suggest that the mean annual TVOC emission for this period was 0.128 Gg, ranging 0.008 Gg to 0.624 Gg. While VOCs emissions varied widely from year to year, they decreased substantially since 2004 with the work of prevention, management of forest areas and less critical conditions. There is a link between the emission of VOCs and the driest seasons. In 2000 and 2003 there were severe droughts and as a consequence big fires occurred that for which maximum VOCs were emitted.

REFERENCES

Adams, RP. Identification of essential oils by capillary gas chromatography/ mass spectroscopy," *Carol Stream*, IL: Allured, 2001.

Andreae, MO; Merlet, P. Emission of trace gases and aerosols from biomass burning, *Global Biogeochemical Cycles*, Vol. 15, 955-966, 2001.

Barboni, T; Chiaramonti, N. BTEXs emissions according to the distance from the flame front and the combustion stages during the prescribed burning, *Combustion Science and Technology*, Vol. 182, 1193–1200, 2010.

Barboni, T; Pelizzaro, G; Arca, B; Chiaramonti, N; Duce, P. Analysis and origins of smoke from ligno-cellulosic fuels, *Journal of Analyical and Applied Pyrolysis*, Vol. 89, 60–65, 2010.

Barboni, T; Cannac, M; Pasqualini, V; Simeoni, A; Leoni, E; Chiaramonti, N. Volatile and semi-volatile organic compounds in smoke exposure of firefighters during prescribed burning in the Mediterranean region, *International Journal of Wildland Fire*, Vol. 19, 606–612, 2010.

Barboni, T; Cannac, M; Leoni, E; Chiaramonti, N. Emission of biogenic volatile organic compounds involved in eruptive fire: implications for the safety of firefighters, *International Journal of Wildland Fire*, Vol. 20, 152–161, 2011.

Barboni, T; Morandini, F; Rossi, L; Molinier, T; Santoni, P-A. Measure of the fireline intensity obtained by calorimetry and the relationship with the flame length, *Combustion Science and Technology*, Vol. 184, 186–204, 2012.

Chetehouna, K; Barboni, T; Zarguili, I; Leoni, E; Simeoni, A; Fernandez Pello, AC. Investigation on the emission of Volatile Organic Compounds from heated vegetation and their potential to cause an eruptive forest fire, *Combustion Science and Technology*, Vol. 181, 1273-1288, 2009.

Dost, FN. Acute Toxicology of Components of Vegetation Smoke, *Reviews of Environmental Contamination and Toxicology*, Vol. 119, 1–46, 1991.

Evtyugina, M; Calvo, AI; Nunes, T; Alves, C; Fernandes, AP; Tarelho, L; Vicente, A; Pio, C. VOC emissions of smouldering combustion from Mediterranean wildfires in central Portugal, *Atmospheric Environment*, Vol. 64, 339-348, 2013.

Evtyugina, M; Alves, C; Calvo, AI; Nunes, T; Tarelho, L; Duarte, M; Prozil, S; Evtugin, D; Pio, C. VOC emissions from residential combustion of Southern and mid-European woods, *Atmospheric Environment*, Vol. 83, 90-98, 2014.

Flannigan, MD; Bergeron, Y; Engelmark, O; Wotton, BM. Future wildfire in circumboreal forests in relation to global warming, *Journal of Vegetation Science*, Vol. 9, 469–476, 1998.

Friedli, HR; Atlas, E; Stroud, VR; Giovanni, L; Campos, T; Radke, LF. Volatile organic trace gases emitted from North American wildfires, *Global Biogeochemical Cycles*, Vol. 15, 435- 452, 2001.

French, NHF; Goovaerts, P; Kaischke, ES. Uncertainty in estimating carbon emissions from boreal forest fires, *Journal of Geophysical Research*, Vol. 109, D14S08, 2004.

Gouw, JA; Warneke, C; Stohl, A; Wollny, AG; Brock, CA; Cooper, OR; Holloway, JS; Trainer, M; Fehsenfeld, FC; Atlas, EL; Stroud, V; Lueb, A. Volatile organic compounds composition of merged and aged forest fire plumes from Alaska and western Canada, *Journal of Geophysical Research: Atmospheres*, Vol. 111, D10303, 2006.

Kasischke, ES; Bruhwiler, LP. Emissions of carbon dioxide, carbon monoxide, and methane from boreal forest fires in 1998, *Journal of Geophysical Research*, Vol. 108, 8146, 2003.

König, WA; Hochmuth, DH; Joulain, J; Terpenoids D; and related constituents of essential oils, library of massfinder 2.1. Hamburg: University of Hamburg, Institute of Organic Chemistry, 2001.

Lü, A; Tian, H; Liu, M; Liu, J; Melillo, JM. Spatial and temporal patterns of carbon emissions from forest fires in China from 1950–2000, *Journal of Geophysical Research*, Vol. 111, D05313, 2006.

Miranda, AI. An integrated numerical system to estimate air quality effects of forest fires, *International Journal of Wildland Fire*, Vol. 13, 217–26, 2004.

Miranda, AI; Martins, V; Cascão, P; Amorim, JH; Valente, J; Tavares, R; Borrego, C; Tchepel, O; Ferreira, AJ; Cordeiro, CR; Viegas, DX; Ribeiro, LM; Pita, LP. Monitoring of firefighters exposure to smoke during fire experiments in Portugal, *Environment International*, Vol. 36, 736–745, 2010.

Narayan, C; Fernandes, PM; Van Brusselen, J; Schuck, A. Potential for CO2 emissions mitigation in Europe through prescribed burning in the context of the Kyoto Protocol, *Forest Ecology and Management*, Vol. 251, 164–173, 2007.

National Institute of Standards and Technology NIST WebBook. <http://webbook.nist.gov/chemistry> 2005.

Proterina-C, Rapport sur les résultats obtenus dans les activités de campagne et leur insertion dans la base de données, rapport interne, In french, 2011.

Reisen, F; Brown, SK. Australian firefighters' exposure to air toxics during bushfire burns of autumn 2005 and 2006, *Environment International*, Vol. 35, 342–352, 2009.

Santoni, P-A; Morandini, F; Barboni, T. Determination of fireline intensity by oxygen consumption calorimetry, *Journal of Thermal Analysis and Calorimetry*, Vol. 104, 1005-1015, 2011.

Seiler, W; Crutzen, PJ. Estimates of gross and net fluxes of carbon between the biosphere and the atmosphere from biomass burning, *Climatic Change*, Vol. 2, 207–247, 1980.

Sinha, P; Hobbs, PV; Yokelson, RJ; Bertschi, IT; Blake, DR; Simpson, IJ; Gao, S; Kirchstetter, TW; Novakov, T. Emissions of trace gases and particles from savanna fires in southern Africa, *Journal of Geophysical Research*, Vol. 108, D8487, 2003.

Sinha, P; Hobbs, PV; Yokelson, RJ; Blake, DR; Simpson, IJ; Gao, S; Kirchstetter, TW. Emissions from miombo woodland and dambo grassland savanna fires Journal of Geophysical Research, Vol. 109, D11305, 2004.

Statheropoulos, M; Karma, S. Complexity and origin of the smoke components as measured near the flame-front of a real forest fire incident: A case study, *Journal of Analyical and Applied Pyrolysis*, Vol. 78, 430–437, 2007.

Stocks, BJ; Fosberg, MA; Lynham, TJ; Mearns, L; Wotton, BM; Yang, Q; Lin, J-Z; Lawrence, K; Hartley, GR; Mason, JA; McKenney, DW. Climate change and forest fire potential in Russian and Canadian boreal forests, *Climatic Change*, Vol. 38, 1–13, 1998.

Stocks, BJ; Mason, JA; Todd, JB; Bosch, EM; Wotton, BM; Amiro, BD; Flannigan, MD; Hirsch, KG; Logan, KA; Martell, DL; Skinner, WR. Large forest fires in Canada, 1959–1997, *Journal of Geophysical Research*, Vol. 108, 1-12, 2003.

Urbanski, J; Yokelson, RJ; Hao, WM; Baker, S. Chemical Composition of Wildland Fire Emissions in: Bytnerowicz, A., Arbaugh, M., Riebau, A., Anderssen, C. (Eds.), Wildland Fires and Air Pollution, *Developments in Environmental Science*, V. 8, Elsevier, Amsterdam, 2008.

US National Institute of Standards and Technology PC Version 1.7 of the NIST/EPA/NIH Mass Spectra Library, Norwalk, CT, 1999.

Yokelson, RJ; Karl, T; Artaxo, P; Blake, DR; Christian, TJ; Griffith, DWT; Guenther, A; Hao, WM. The tropical forest and fire emissions experiment: overview and airborne fire emission factor measurements, *Atmospheric Chemistry and Physics*, Vol. 7, 5175-5196, 2007.

Yokelson, RJ; Christian, TJ; Karl, TG; Guenther, A. The tropical forest and fire emissions experiment: laboratory fire measurements and synthesis of campaign data, *Atmospheric Chemistry and Physic*, Vol. 8, 4221–4266, 2008.

Ward, DE. Smoke from wildland fires. In: Goh K-T, Schwela D, Goldammer JG, Simpson O, editors. Health guidelines for vegetation fire events— background papers, Lima, Peru, Paper 70–85, 1998.

In: Volatile Organic Compounds
Editor: Khaled Chetehouna

ISBN: 978-1-63117-862-7
© 2014 Nova Science Publishers, Inc.

Chapter 3

BIOGENIC VOLATILE ORGANIC COMPOUNDS EMISSIONS OF HEATED MEDITERRANEAN VEGETAL SPECIES

K. Chetehouna[1], L. Courty[2], L. Lemée[3], F. Bey[1] and J. P. Garo[4]*

[1]INSA-CVL, Univ. Orléans, PRISME EA 4229, F-18022, Bourges, France
[2]Univ. Orléans, INSA-CVL, PRISME EA 4229, F-18000, Bourges, France
[3]Univ. de Poitiers, CNRS UMR 7285 (IC2MP), F-86073, Poitiers, France
[4]Institut P', UPR 3346 CNRS, ENSMA, Univ. Poitiers, F-86961, Futuroscope Chasseneuil, France

ABSTRACT

The effect of temperature on Biogenic Volatile Organic Compounds (BVOCs) emissions from five vegetal species was studied between 70 and 180 °C, range during which vegetation produces high amount of these gases. Emissions were investigated at small and middle scales. For a given species, if emitted BVOCs content does not change, the relative percentage of each volatile compound varies according to temperature. High amount of terpenoid compounds were emitted, except for *Cistus albidus*, and emissions are increasing with temperature. The main identified compounds are thymol, 1-fenchone, α-pinene, 3-hexen-1-ol and limonene for respectively *Thymus vulgaris*, *Lavandula stœchas*,

---

* Corresponding author: Email: khaled.chetehouna@insa-cvl.fr.

Rosmarinus officinalis, *Cistus albidus* and *Pinus pinea*. The results obtained will permit to develop a database of BVOCs emissions at elevated temperature and to include their combustion in physical forest fires models.

Keywords: BVOCs, Emissions scales, Flash pyrolysis, *Thymus vulgaris, Lavandula stœchas, Rosmarinus officinalis, Cistus albidus, Pinus pinea*

INTRODUCTION

Almost all vegetal species produce (store and emit) Biogenic Volatile Organic Compounds (BVOCs), principally in their leaves. These compounds are nitrogen-free hydrocarbons, very reactive and they are part of secondary metabolites of vegetal. Volatile Organic Compounds (VOCs) have effects on human health and many of them are even carcinogen. Indeed, many studies have correlated the presence of VOCs to the occurrence of respiratory diseases in a given population (Rumchev et al. 2004, Arif et Shah 2007, Bernstein et al. 2008, Billionnet et al. 2011). They also play an important role towards the environment, especially in the atmospheric pollution processes. In summer, amounts of biogenic VOCs are higher than amounts of VOCs from anthropic origin (Simon et al. 2001). The major impact of VOCs from plant origin on the atmosphere chemistry is observed on the processes of creation and degradation of the tropospheric nitrogen at low altitude. Despite the important role of biogenic VOCs in atmospheric pollution, it is also necessary to take into account these compounds in the ignition risk of vegetation (Owens et al. 1998, Nuñez-Regueira et al. 2005, Ormeño et al. 2009), particularly in forest fires since these compounds have very low values of Lower Flammability Limit (LFL) and flashpoint. For instance, limonene (99 % pure) has a LFL of 0.7 % vol. in air and flashpoint of 43 °C. VOCs can therefore lead to a modification of wildland fires dynamic. Indeed, it has been reported that in some occasions fires can behave in a surprising way, suddenly changing from a moderate behavior (characterized by a relatively low rate of spread) to an explosive propagation characterized by higher energy released and spread rate . This phenomenon is known as Accelerating Forest Fire (AFF). The origin of such a phenomenon can be the accumulation of biogenic VOCs, reaching concentrations closed to the BVOCs/Air mixtures LFL during seasons where the plants are very flammable (Chetehouna et al. 2009, Viegas and Simeoni 2011, Courty et al. 2012).

Wildland fires affect the natural environment by damaging different vital parts of the vegetation (foliage, stem and roots) and they can also have effects on phyto-sanitary risks, on the regeneration of forest stand, on soil, on vegetation dynamics, on fauna, on the landscape (International Handbook on Forest Fire Protection). Concerning their effects on the soil, forest fires lead to a reduction of the water retention capacity and of the rate of water infiltration because the soil becomes less porous. The soil composition is affected because volatile minerals are involved in the convection column in the form of very fine particles, and are displaced several kilometers away from their initial place. Moreover, vegetation recovery depends on the forest fire frequency and severity. Although in almost all cases vegetation recovers quickly without any human intervention, three cases can be distinguished. After moderate forest fire, vegetation gradually recovers by resprouting, germination, or starting from underground parts that survived. An intense forest fire reduces the regeneration capacities and it results in a floristic impoverishment. Frequently occurring fires lead to a significant floristic impoverishment and many vegetal species do not have enough time to mature for sexual reproduction before the occurence of another forest fire. Gases emitted by the combustion process can react with ultraviolet rays emitted by the sun and therefore create a photochemical pollution. Smokes contain polycyclic aromatic hydrocarbons and tars that contribute to air pollution. High amount of carbon dioxide, which is one of the main greenhouse effect gases, is emitted thus participating to global warming.

We can find several works in the literature dealing with BVOCs emissions from various vegetal species at room temperature. Indeed, Owen et al. (2001) studied the emissions of forty Mediterranean plant species and observed thirty two compounds. Ormeño et al. (2007) investigated the emissions of three typical Mediterranean plants at atmospheric conditions. The BVOCs emissions of *Pinus pinea* needles, branches and female cones were analyzed by Macchioni et al. (2003) at ambient temperature and it demonstrated that limonene is the major constituent emitted by these three plant parts. There is a lack in the literature on the BVOCs emissions of plants as functions of temperature, which is essential for forest fire models because large amounts of BVOCs are emitted for temperatures higher than the ambient temperature. We can note the studies of Chetehouna et al. (2009) and Barboni et al. (2011) for respectively the emissions of *Rosmarinus officinalis* in an hermetic enclosure and the emissions of *Pinus laricio*, *Pinus pinaster* and *Cistus monspeliensis* using an Automatic Thermal Desorber (ATD). This chapter deals with the emissions of BVOCs of five heated Mediterranean vegetal species.

BVOCs are generally sampled at three scales: (i) from leaves or needles (i.e., small scale, Isidorov et al. 2003, Greenberg et al. 2006), (ii) from plants grown in greenhouses (i.e., midle scale, Llusià et Peñuelas 1999, Pegoraro et al. 2004) and (iii) from trees in their natural environment (i.e., terrain scale, Cooke et al. 2001, Moukhtar et al. 2006). The main purpose of this chapter is to characterize the emissions of BVOCs from five heated Mediterranean species in order to extend the existing literature database on plants BVOCs emissions. The five plant species selected for this work and typical of the Mediterranean basin are *Rosmarinus officinalis*, *Pinus pinea*, *Thymus vulgaris*, *Lavandula stoechas* and *Cistus albidus*. The first one was studied at both midle and small scales, whereas the four others were only investigated at the leave or needle scale. The second section is devoted to the description of the experimental setups for BVOCs emissions and to the optimisation of the experimental parameters. The emissions of these five plant species as function of preheat temperature are presented and discussed in the third section. The effect of the fire retardant on the BVOCs emission are at least investigated at the plant scale.

METHODS

Experimental Methodology at Middle Scale

Fifty *Rosmarinus officinalis* shrubs were placed in a hermetic enclosure and heated by a radiant panel. These plants were used to determine the effects of plant temperature and fire retardant on the production of BVOCs. These experiments were conducted at different periods of time with different ambient conditions. The hermetic enclosure (100 cm × 100 cm × 134 cm) was manufactured in Siporex material (thickness of 7 cm) and its volume was about 1.2 m^3. The radiant panel was an assembly of 16 black ceramics plates of 12 x 12 cm^2 providing a thermal power of 1.2 kW per plate. The maximal power emitted per unit of surface of this radiant panel was about 83 $kW.m^{-2}$ (Figure 1). The plants of *Rosmarinus officinalis* of an average mass of 155 g and an average height of 30 cm were placed at the center of the hermetic enclosure at a distance of 50 cm from the radiant panel. The moisture content of the different plants is about 70 % and they were heated during 30 min. The heat flux of the radiant panel varied from 0.44 to 20.59 $kW.m^{-2}$ for the study of the plant temperature (or heat flux) effect and was fixed at 15.50 $kW.m^{-2}$ to study the effect of the quantity of the fire retardant. The trapping and the sampling of the BVOC were ensured by sorbent tubes and a pump at a flow-rate of 150 mL.min^{-1}. The sorbent tubes used were glass multibed tubes (11.5 cm x 6 mm o.d. x 4 mm i.d), Supelco® (Tenax TA®). These tubes were transferred in a freezing box into the laboratory and analysed using ATD-GC/MS. The sampling was carried out in triplicate. The experimental protocol is illustrated in Figure 2. Each sampling and analysis was performed in triplicate.

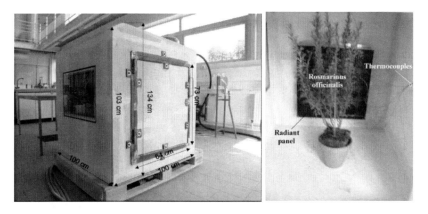

Figure 1. Pictures of the hermetic enclosure, of the plant and of the radiant panel.

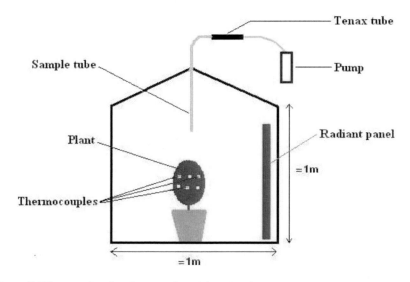

Figure 2. Diagram showing the experimental protocol.

The analyses were carried out using an Automatic Thermal Desorber Perkin Elmer® (Waltham, MA, USA) ATD 400 turbomatrix®. For the thermal desorption of volatile components, helium flow was set at 30 mL. min[-1]. A sorbent tube (Tenax TA®) was brought at 280 °C for 10 min and a carrier gas (helium) flushed the sample towards a cold trap (Tenax TA®, 22 cm, 0.53 mm i.d., Supelco® Co.) at -30 °C. In a second step, the cold trap was programmed from -30 to 280 °C at 40 °C/s then held isothermally at 280 °C for 3 min. The compounds were desorbed to the column of the GC/FID and column of the GC/MS by the transfer line maintained at 280 °C. GC/FID analyses were carried out using a Perkin-Elmer® apparatus equipped with a Flame Ionization Detector (FID) and a fused-silica capillary column RTX-1 (polydimethylsiloxane, 60 m × 0.22 mm i.d., film thickness 0.25 μm). The oven temperature was programmed from 60 °C to 230 °C at 2 °C/min and then held isothermally at 230 °C for 35 min. Retention indices (I) of compounds were determined relatively to the retention times of series of n-alkanes (C5-C30) with linear interpolation, using Van den Dool and Kratz equation (Van den Dool and Kratz, 1963), and software from Perkin-Elmer®. GC/MS analyses were carried out using Perkin-Elmer® apparatus equipped with a TurboMass® detector (quadrupole), and a fused-silica capillary column RTX-1 (polydimethylsiloxane, 60 m × 0.22 mm i.d., film thickness 0.25 μm). Ion source temperature was set at 150 °C, Electron impact (EI) ionization at 70 eV and EI mass spectra were acquired over the mass range 35-350 Da (scan time: 1s). The oven temperature was programmed from 60 °C to 230 °C at 2 °C/min and then held isothermally at 230 °C for 35 min. The methodology carried out for identification of individual components was based:

1. on computer matching with commercial mass spectral libraries (NIST/EPA/NIH, 1999), and comparison of mass spectra with those of our own library of authentic compounds or literature data (König et al., 2001).
2. on comparison of calculated retention indices on apolar column, with those of authentic compounds or literature data (NIST WebBook, 2005).

The quantification of the different compounds is presented in α-pinene equivalent. The quantification of this compound was carried out by direct liquid injection in the Tenax tubes and was calibrated in mass range 3.58 ng-17.16 μg.

Experimental Methodology at Small Scale

Flash pyrolysis technique has been used by several authors to study the degradation of different types of materials (Page et al. 2002, Sezer et al. 2008, Gascoin et al. 2013). This technique was applied in this work on five vegetal species using a CDS Pyroprobe 5150 equipped with a 5250 autosampler. Between 2.3 and 7.3 mg (according to the species) of needle samples were introduced into a 2 i.d. × 40mm quartz tube and inserted into the autosampler where they were heated up to the desired temperature. The temperature rise can be varied up to a maximum heating rate of 5000 °C.s^{-1}.The emitted BVOCs were carried out by helium via a transfer line heated at 280 °C to the GC/MS.

The latter is a ThermoFisher Focus gas chromatograph (GC) coupled with a DSQ II quadripole mass spectrometer (MS). GC separations were conducted using a BPX (SGE) capillary column (30m long, 0.25mm i.d. and 0.25 μm phase thickness). The column temperature was programmed from 40 to 200 °C at a rate of 5 °C.min^{-1} and held for 5 minutes at 200 °C. Mass spectra were recorded in the electron impact mode with ionization energy of 70 eV. Identification of BVOCs is based on a comparison of their mass spectra with the NIST mass spectral library, with data from literature and with mass spectra and retention times of reference compounds (α-pinene, limonene). These reference compounds were also used to make calibration curves for quantification.

The selected temperature range was between 70 and 180 °C in order to determine BVOCs emissions before their pyrolysis phase (Granström 2003). Three experiments were performed for each temperature in order to ensure the repeatability. A schematic overview of the emission experimental setup at small scale is illustrated in Figure 3.

Raffali et al. (2002) studied the emissions of *Rosmarinus officinalis* needles as a function of temperature and showed that large amounts of BVOCs were emitted at about 180 °C. For this reason, we have selected this temperature value to study the effects of two parameters: the heating time and heating rate. Two approaches were followed, one varying the heating time from 20 to 60 s for a constant heating rate of 5000 °C/s and the other varying the heating rate from 1 to 5000 °C/s for a constant heating time of 30 s. Experiments were performed with *Thymus vulgaris* needles. 13 compounds were identified and these BVOCs are the same for different values of heating time and heating rate.

Figures 4 and 5 show respectively the influence of the heating time and heating rate on the emissions of BVOCs by *Thymus vulgaris* species. It is seen from Figure 4 that three major compounds, namely thymol ($C_{10}H_{14}O$), p-cymene ($C_{10}H_{14}$) and γ-terpinene ($C_{10}H_{16}$) are the same for the four heating times. Moreover, they represent more than 85 % of the mixture for each heating times. It can also be noticed that the amounts emitted for each compound are the largest for a heating time of 30 s indicating that this value can be chosen to study the effect of temperature on the BVOCs emissions. From Figure 5 it can be seen that the same three major compounds are identified for different heating times, and for all the heating rates. They represent more than 70 % of the mixture for the five studied heating rates. It is also noticed that the evolution of the BVOCs relative emission has a non-monotonic behavior with a maximum emission for 5000 °C/s. This heating rate is the maximum value that can be provided by the flash pyrolysis apparatus (CDS Pyroprobe 5150) and was selected for the study of the temperature effect on the BVOCs emissions. Emissions experiments were performed at small scale with these adapted parameters. The studied vegetal species were *Thymus vulgaris*, *Rosmarinus officinalis*, *Lavandula stœchas*, *Cistus albidus* and *Pinus pinea*.

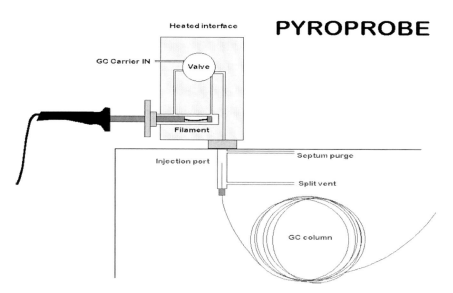

Figure 3. Schematic overview of the emission experimental setup at small scale.

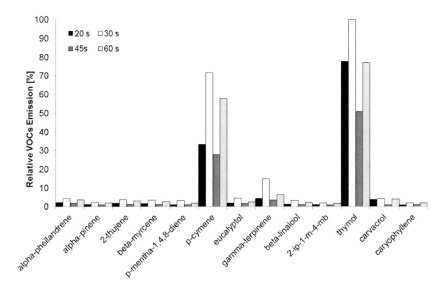

Figure 4. BVOCs emissions of Thymus vulgaris needles at 180 °C for different heating times.

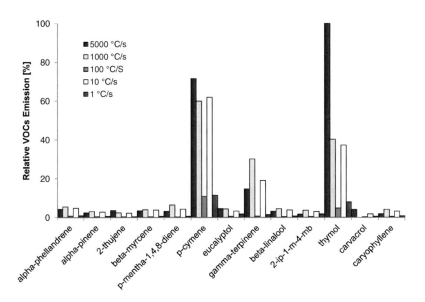

Figure 5. BVOCs emissions of *Thymus vulgaris* needles at 180 °C for different heating rates.

RESULTS AND DISCUSSION

Emissions at Middle Scale

Effect of Plant Temperature and Radiant Panel Heat Flux

The aim of these tests is to track the BVOCs emission as a function of temperature in order to estimate the quantity emitted by vegetation during a forest fire. As indicated previously, we chose a radiant panel heat flux in the range 0.44 to 20.59 $kW.m^{-2}$ corresponding to a temperature range between 30 and 210 °C, a sampling height of 112 cm and heating and sampling times equal to 30 min and 10 min respectively. The BVOCs analyses were carried out using ATD-GC/MS, 18 compounds were identified for the *Rosmarinus officinalis* (Lamiaceae family) emissions and classified by ascendant order of the retention index I (cf. Table 1).

Table 1. Identified VOCs emitted by *Rosmarinus officinalis*

$N°$	^1I_l	^2I_a	Name
☐	936	931	α-pinene
2	950	943	camphene
3	964	952	thuja-2,4-(10)-diene
4	970	964	sabinene
5	978	970	β-pinene
6	982	979	myrcene
7	1013	1008	α-terpinene
8	1018	1012	para-cymene
9	1022	1020	1,8-cineol
10	1025	1020	limonene
11	1051	1047	γ-terpinene
12	1082	1079	α-terpinolene
13	1088	1085	filifolone
14	1128	1122	camphor
15	1149	1141	pinocarvone
16	1182	1176	estragol
17	1219	1210	verbenone
18	1270	1269	bornyl acetate

[1] retention index from literature data (NIST WebBook, 2006).
[2] retention index calculated on apolar column.

The BVOCs composition was characterized by a high content of monoterpenic hydrocarbons. The main components were α-pinene, limonene, camphene, myrcene, β-pinene, para-cymene and camphor. The same constituents were observed by Ormeño et al. (2007) for the study of the BVOCs emissions by *Rosmarinus officinalis* under natural conditions. The total emission of these compounds represents a percentage varying with temperature from 88 to 100% of the total BVOCs emissions. Figure 6 describes the evolution of the main constituents and of the total BVOCs emissions as a function of the enclosure temperature. It was also observed that, when the temperature increases, the BVOCs emission increases until reaching a temperature of 175 °C. This tendency was observed by Barboni et al. (2011) for others Mediterranean vegetal species such as *Pinus nigra* ssp *laricio* and *Pinus pinaster*. The BVOCs quantity at 175 °C is 8 times higher than the one measured at 50 °C. α-pinene is the major compound identified in *Rosmarinus officinalis* emissions. α-Pinene emissions are five times higher at 175 °C than at 50 °C. Moreover, we observe an increase of BVOCs emission around 120 °C due to the transport of the BVOCs by the water molecules during their evaporation process. Between the temperatures 120 °C and 150 °C, we observe a diminution of the BVOCs quantity probably due to the lower influence of the water evaporation. Since the boiling temperature of monoterpenes is around 154 °C, for the temperature below 150 °C these molecules would be in a liquid or in an equilibrium liquid-vapour state. Consequently, for temperatures higher than 150 °C, BVOCs emissions increase rapidly to reach a maximum at 175 °C. The amount of volatile compounds is 3.3 times more important in this temperature range than in the 50 °C-120 °C range. Beyond 175 °C, we notice a significant decrease of the quantity of biogenic VOCs that can be explained by the thermal degradation and/or polymerisation of the terpenes.

Fire Retardant Effect

A great part of the fight against forest fires consists in dropping liquids from aircrafts. The dropped fluid can be water for direct attack or chemical retardant, which is in Europe a mixture of water (80%), polyphosphate, clay and gum. Several studies have been carried out for the effect of fire retardants on the forest fires behaviour (Giménez et al. 2006, Liodakis et al. 2006), but there are no works devoted to their effect on BVOCs emission. In this work we conducted a study on the effect of applying different quantities of chemical fire retardant (η_1=6 mL, η_2=12 mL, η_3=18 mL, η_4=24 mL and η_5=30 mL) on BVOCs emission. We also used water (30 mL) in order to compare its effects to the fire retardant effects on the BVOCs emission. The enclosure temperature in this study corresponds to the maximum value of the BVOCs emission previously found (175 °C). The total emission of the seven main compounds presented above represents a percentage varying with the chemical fire retardant quantity from 92 to 98% of the total BVOCs emissions. The different results are presented in Figure 7. It is observed reading this Figure that the application of fire retardant favors the BVOCs emission. The total quantity of BVOCs changes from the value of 0.0424 µg.g$^{-1}_{dw}$ without fire retardant to the value of 0.0522 µg.g$^{-1}_{dw}$ with quantity of η_1=6 mL. Moreover, the BVOCs emission increases while increasing the quantity of fire retardant and it reaches a value of 0.1196 µg.g$^{-1}_{dw}$ for the saturation quantity of η_5=30 mL. This result can be explained by the presence of water, which causes the transport of the BVOCs molecules as it evaporates. Furthermore, the comparison between water and fire retardant mixture and water alone, the two cases with quantity of η_5=30 mL, shows that the total emissions of BVOCs are about 2 times more important with water alone. Thus, the chemical fire retardant has a reducing effect on the BVOCs emission. This reduction is due to the presence of clay in this substance which plays an adsorbent role as indicated by Harti et al. (2007) and Zaitan et al. (2008).

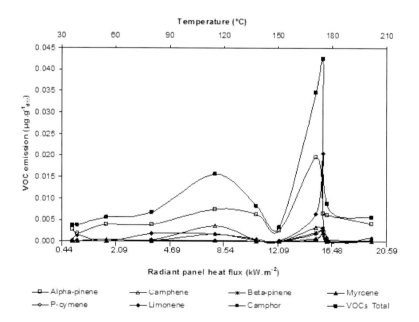

Figure 6. BVOCs emission as a function of temperature and radiant heat flux.

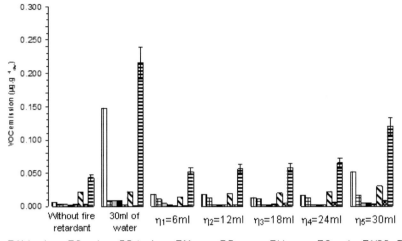

Figure 7. Effect of the fire retardant on the BVOCs emissions at 175 °C.

Effect of Radiant Panel Heat Flux, Plant Temperature and Fire Retardant

As showed in the previous section, the application of the fire retardant increases the BVOCs emission due to the presence of water (80%), and for the same quantity the fire retardant reduces considerably the BVOCs emissions (about 55%). To explain this result, we have analysed the behaviour of the BVOCs emissions as a function of temperature with the presence of the fire retardant. We chose the maximal quantity of $\eta_5=30$ mL and a radiant panel heat flux range from 0.44 to 20.59 kW.m^{-2}, corresponding to a temperature range from 30 to 210 °C. The different results are presented in Figures 8 and 9. We observe that the total emissions of the seven main compounds, namely α-pinene, limonene, camphene, myrcene, β-pinene, p-cymene and camphor, varie with temperature from 90 to 98% of the total BVOCs emissions. Figure 8 shows an increase in the BVOCs emissions as the temperature increased to a peak around 180°C and a decrease above this temperature. As explained above, the BVOCs emission is favoured by the water evaporation. From 154 °C that corresponds to the boiling point of BVOCs there is a significant increase in emissions due to their evaporation, until a maximum value at 180 °C. Above this threshold temperature, the decrease in BVOCs emissions seems to be due to the thermal degradation and/or polymerisation of terpenic compounds. The comparison of the variation of total BVOCs emissions with and without the fire retardant versus the plant temperature and radiant panel heat flux (cf. Figure 8) shows a similar trend. Furthermore, the total quantity of BVOCs with the fire retardant is higher than the one without and it is 3 to 5 times more important. As indicated above, this result is explained by the presence of a significant quantity of water, which enhances the BVOCs emissions through its evaporation.

Emissions at Small Scale

Thymus Vulgaris Emissions

Thirteen compounds were identified for *Thymus vulgaris*, whatever the initial temperature is, and three major compounds appear: thymol, p-cymène and γ-terpinene. Temperature has no influence on the emitted BVOCs but on their amounts and relative proportions in the mixture. Table 2 presents the thirteen compounds emitted by *Thymus vulgaris*, giving their names, retention times, chemical formulas, chemical families and skeletal formulas.

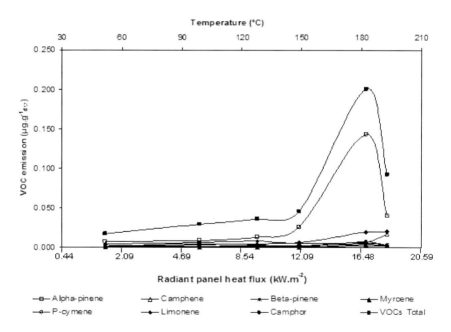

Figure 8. Evolution of VOCs emission as a function of temperature and heat flux with the presence of fire retardant (η5=30mL).

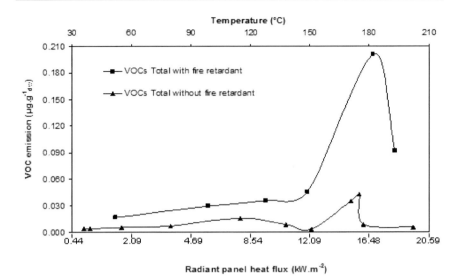

Figure 9. The total BVOCs emission as a function of plant temperature and radiant heat flux with and without fire retardant (η5=30 mL).

The identified BVOCs can be divided into mono and sesquiterpenes. Amongst monoterpenic compounds we observed 4 families: monoterpenes, monoterpenic alcohols, ethers and phenols. The relative amount of the two major compounds, thymol and p-cymene, are presented in Figure 10 and compared with the total BVOCs. Emissions X are quantified taking into account the peak area A_i of each compound and the mass m of the pyrolysed sample: $X = \frac{A_i}{m}$. Values presented here (X_n) are normalized on the highest value (i.e., emissions of total BVOCs at 180 °C): $X_n = \frac{X}{X_{max}} \cdot 100$. Figure 10 shows that the concentration of emitted BVOCs is correlated with temperature and become significant above 125 °C. The increase is exponential since emitted BVOCs amount at 180 °C is one thousand times higher than the amount measured at 70 °C, whereas the amount of thymol is ten thousand times more important at 180 °C than at 70 °C. Total BVOCs amount is multiplied by 220 between 125 °C and 180 °C and by 14 between 150 and 180 °C. Emissions are maximal and very important at 180 °C.

Table 2. BVOCs emitted by *Thymus vulgaris*: name, retention time, chemical formula, chemical family and skeletal formula

BVOC	Retention time [min]	Chemical formula	Chemical family	Skeletal formula
α-phellandrene	8.72	$C_{10}H_{16}$	Monoterpene	
α-pinene	8.96	$C_{10}H_{16}$	Monoterpene	
2-thujene	10.38	$C_{10}H_{16}$	Monoterpene	
myrcene	10.55	$C_{10}H_{16}$	Monoterpene	
terpinolene	11.45	$C_{10}H_{16}$	Monoterpene	
p-cymene	11.75	$C_{10}H_{14}$	Monoterpene	
cineol	11.96	$C_{10}H_{18}O$	Monoterpenic ether	
γ-terpinene	12.70	$C_{10}H_{16}$	Monoterpene	

BVOC	Retention time [min]	Chemical formula	Chemical family	Skeletal formula
β-linalool	13.95	$C_{10}H_{18}O$	Monoterpenic alcohol	
isopropyl-methoxy-methylbenzene	18.03	$C_{11}H_{16}O$	Monoterpenic ether	
thymol	19.68	$C_{10}H_{14}O$	Monoterpenic phenol	
carvacrol	19.86	$C_{10}H_{14}O$	Monoterpenic phenol	
β-caryophyllene	22.97	$C_{15}H_{24}$	Sesquiterpene	

Figure 11 illustrates the composition of BVOCs mixtures emitted by *Thymus vulgaris* needles at different temperatures. It is interesting to note that thymol represents more than 50 % of the total emissions at temperatures where high amount of BVOCs are emitted (i.e., above 125 °C). The maximal percentage of thymol is observed at 150 °C, where it represents more than 72 % of the mixture. Thymol is the main compound for all temperatures, even if it represents only 24 % of the mixture at 90 °C. The percentage of the second compound, *p*-cymene, is more constant and is between 15 and 25 % for all temperatures. Except these two major compounds, other BVOCs represent more than 40 % of the mixture for temperatures lower than 125 °C; above this temperature, they don't represent more than 18 % of the total BVOCs. Nezhadali et al. (2010) and Parra et al. (2004) studied the emissions of *Thymus vulgaris* under natural conditions. Parra et al. (2004) estimated the magnitude of non-methane volatile organic compounds emitted by vegetation in Catalonia, Spain, along with their superficial and temporal distribution. They first defined mathematically an emission factor and then found that for *Thymus vulgaris* it was lower for monoterpenes than for other BVOCs. This result is in agreement with our study. In their paper, Nezhadali et al. (2010) compared the composition of *Thymus vulgaris* emissions using two different extraction methods: hydrodistillation and headspace solid phase microextraction. They found that for the 2 methods, the major compounds were thymol, p-cymene, γ-terpinene, myrcene, α-pinene and caryophyllene. They found similar results for the two studied methods but concluded that hydrodistillation needs more time and much more amount of plant. They studied the emissions at 25 and 50 °C and obtained results that are very similar to those of the present study, even if they only work at low temperatures. Another study about the natural emissions of *Thymus vulgaris* is the one of Owen et al. (2001). These authors studied the BVOCs emitted from 40 Mediterranean plant species. They found that the two major compounds for *Thymus vulgaris* were p-cymene and thymol; the following ones were α-pinene and β-myrcene. However, there are slight differences between their results and the present ones, especially for the major compound. It can be explained by the differences in the plant characteristics and in the experimental protocol. Indeed, they used plants collected in a natural environment whereas in this study they were grown in a greenhouse. They also used a sampling technique called Teflon branch enclosure that is very efficient for natural conditions but are not appropriate to simulate the conditions of an ongoing forest fire.

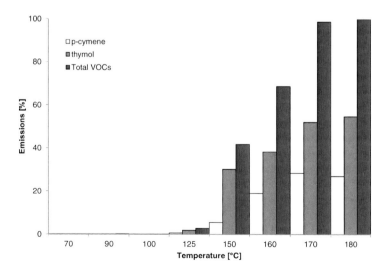

Figure 10. BVOCs relative emissions of *Thymus vulgaris* needles at different temperatures.

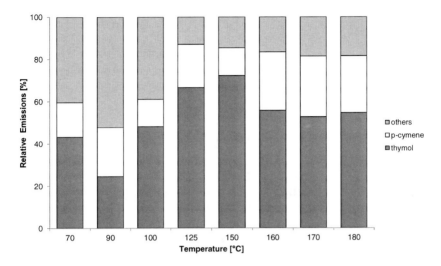

Figure 11. Composition of the BVOCs mixtures emitted by Thymus vulgaris needles at different temperatures.

Lavandula Stœchas Emissions

Fourteen compounds have been identified for this vegetal species, whatever the initial temperature is. 1-Fenchone ($C_{10}H_{16}O$) and camphor ($C_{10}H_{16}O$) are the major components. Table 3 presents the fourteen BVOCs emitted by *Lavandula stœchas*, stating their names, retention times, chemical and skeletal formulas and chemical family. This table shows that different types of compounds are emitted by *Lavandula stœchas*, with few monoterpenes ($C_{10}H_{16}$ compounds, only three compounds). Six different families are identified for compounds having six different chemical formulas. The major compound is 1-fenchone and its evolution as a function of temperature is presented in Figure 12, along with the evolution of the total BVOCs. It can clearly be seen in this figure that emissions increase with temperature, both emissions of 1-fenchone and total BVOCs. They present proportional trends, 1-fenchone representing 35 % of the mixture in average (this will be illustrated in the next figure). Emitted amounts become important from 125 °C and it can be noticed a sudden increase between 100 and 125 °C, where the amounts are multiplied by four. The increase is also much more important between 170 and 180 °C (amounts emitted multiplied by two between these two temperatures) than between 150 and 170 °C (multiplied by 1.1). The second major compound identified for most of temperatures is cineol and the composition of the mixtures (two main compounds plus sum of the other BVOCs) for each temperature is presented in Figure 13. The first thing that can be noticed from this Figure is the constancy of each compound (or group of compounds). Indeed, whatever the emitted amount of gases (and therefore whatever the temperature is), l-fenchone represents around 35 %, cineol around 20 % and other BVOCs around 45 %. This tendency is different from the one of other plant species (see comparison section). The two main compounds always represent more than half of the total BVOCs. If we also add camphor to them, the three major BVOCs emitted represent almost 80 % of the mixture.

Table 3. BVOCs emitted by Lavadula stœchas: name, retention time, chemical formula, chemical family and skeletal formula

BVOC	Retention time [min]	Chemical formula	Chemical family	Skeletal formula
Tricyclene	8.65	$C_{10}H_{16}$	Monoterpene	
α-pinene	8.96	$C_{10}H_{16}$	Monoterpene	
Camphene	9.49	$C_{10}H_{16}$	Monoterpene	
p-cymene	11.75	$C_{10}H_{14}$	Monoterpene	
Cineol	11.96	$C_{10}H_{18}O$	Monoterpenic ether	
cis-linaloloxide	13.11	$C_{10}H_{18}O_2$	Epoxide of a monoterpenic alcohol	
α-methyl-4-methyl-3-	13.60	$C_{10}H_{18}O_2$	Epoxide of a monoterpenic	

BVOC	Retention time [min]	Chemical formula	Chemical family	Skeletal formula
pentenyloxirane methanol			alcohol	
l-fenchone	13.80	$C_{10}H_{16}O$	Monoterpenic ketone	
β-linalool	13.95	$C_{10}H_{18}O$	Monoterpenic alcohol	
fenchol-exo	14.74	$C_{10}H_{18}O$	Monoterpenic ether	
Camphor	15.63	$C_{10}H_{16}O$	Monoterpenic ketone	
Borneol	16.32	$C_{10}H_{18}O$	Monoterpenic alcohol	
Verbenone	17.41	$C_{10}H_{14}O$	Monoterpenic ketone	
Thymol	19.68	$C_{10}H_{14}O$	Monoterpenic phenol	

Similar compounds have been identified in essential oil of lavender by Hassiotis and Dina (2010). They found that l-fenchone is the main component, followed by camphor and cineol. This slight difference in the composition can be explained by differences in the temperatures. Indeed, as previously stated, camphor is more important than cineol at low temperatures. L-fenchone has also been identified as major compound in essential oil of *Lavandula stœchas* by Vokou et al. (2002). Strong proportion of camphor in essential oil of lavender has also been noticed by Yassaa and Williams (2005).

Owen et al. (2001) also studied the emissions of flowers and leaves of *Lavandula stœchas* under natural conditions and found that the major compounds are cineol for leaves and camphor for flowers. Second major compounds are cineol for flowers and camphor for leaves. These slight differences with our results can be explained by differences in sampling methods and in the characteristics of the studied plants. Indeed, these authors worked with plants that grew outside and with a specific sampling technique using Teflon branch enclosure.

Rosmarinus Officinalis Emissions

Eighteen compounds have been identified for the emissions of *Rosmarinus officilalis* and the main component is α-pinene. Figure 14 presents the total BVOCs emissions as well as those of α-pinene at different preheat temperatures. It can be noticed from this Figure that the BVOCs emissions increase with needles temperature until 165 °C. As mentioned in the section emissions at middle scale, the same tendency has been observed by Barboni et al. (2011) for other Mediterranean species such as leaves of *Pinus nigra* and *Pinus pinaster*. Moreover, we can see an increase of BVOCs production around 130 °C due to the transport of BVOCs by evaporation process. Knowing that the lowest boiling point of the identified compounds is that of α-pinene and is about 150 °C (Zhao et al. 2011), these molecules would be in a liquid or in a vapour-liquid equilibrium state below this temperature. As a consequence, for the temperatures higher than this value, the BVOCs emission increases rapidly to a maximum at 165 °C. After this temperature, we can observe a diminution of the BVOCs amount that can be explained by the thermal degradation of the emitted terpenoids. The same BVOCs are obtained for all preheat temperatures and an example of the mixture composition at 165 °C (maximum BVOCs emissions) is given in Table 4.

As we can see from this Table, the BVOCs mixture is characterized by high contents of α-pinene, cineole, verbenone, camphor, borneol, β-myrcene and camphene. The total emissions of these compounds represent more than 89% of the total VOCs emissions. Similar constituents are obtained by Ormeño et al. (2007) and Owen et al. (2002) for the study of the BVOCs emissions of *Rosmarinus officinalis* plants at ambient temperature. The emitted compounds can be classified into five chemical families according to their molecular formula. Figure 15 illustrates the repartition of these five groups in the mixture at 165 °C. This Figure exhibits that the BVOCs composition is characterized by a high monoterpenes ($C_{10}H_{16}$) percentage of 45%. These hydrocarbons and the $C_{10}H_{18}O$ compounds represent three-quarters of the total BVOCs emissions. Sesquiterpene hydrocarbons ($C_{15}H_{24}$) are a minority and represent only 1% of the total mixture.

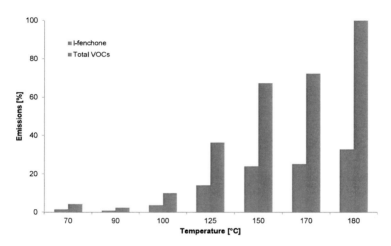

Figure 12. BVOCs relative emissions of Lavandula stœchas leaves at different temperatures.

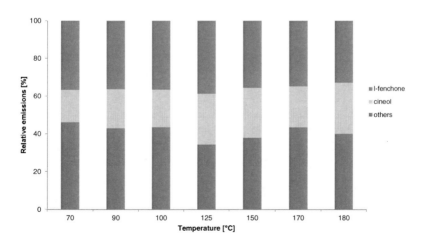

Figure 13. Composition of the BVOCs mixtures emitted by Lavandula stœchas leaves at different temperatures.

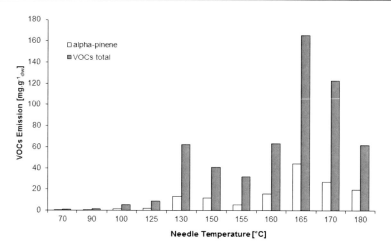

Figure 14. α-pinene and total BVOCs emissions of heated Rosmarinus officinalis needles at different preheat temperature.

Table 4. Emissions of BVOCs from heated *Rosmarinus officinalis* needles at 165 °C

Compound	Molecular formula	Emission ratio (mg.g$_{dw}^{-1}$)	Proportion in the VOCs mixture (%)
α-pinene	$C_{10}H_{16}$	44.401	26.82
camphene	$C_{10}H_{16}$	10.905	6.59
β-myrcene	$C_{10}H_{16}$	13.691	8.27
limonene	$C_{10}H_{16}$	4.155	2.51
cineole	$C_{10}H_{18}O$	27.410	16.55
γ-terpinene	$C_{10}H_{16}$	0.519	0.31
terpinolene	$C_{10}H_{16}$	0.737	0.45
linalol	$C_{10}H_{18}O$	3.302	1.99
chrysanthenone	$C_{10}H_{14}O$	0.913	0.55
camphor	$C_{10}H_{16}O$	15.533	9.38
3-pinanone	$C_{10}H_{16}O$	1.292	0.78
borneol	$C_{10}H_{18}O$	14.387	8.69
terpineol	$C_{10}H_{18}O$	2.926	1.77
verbenone	$C_{10}H_{14}O$	21.381	12.91
geraniol	$C_{10}H_{18}O$	0.798	0.48
terpinen-4-ol	$C_{10}H_{18}O$	0.975	0.59
caryophyllene	$C_{15}H_{24}$	1.984	1.20
α-selinene	$C_{15}H_{24}$	0.272	0.16

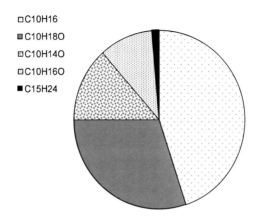

Figure 15. Different molecule groups in the total emissions of heated Rosmarinus officinalis needles at 165 °C.

Cistus Albidus Emissions

Fifteen compounds have been identified for the emissions of *Cistus albidus*, they are presented in Table 5. From this table it is seen that the majority of emitted compounds are sesquiterpenes: eleven compounds on fifteen emitted belong to this family. The major compound and the second one are respectively 3-hexen-1-ol and curcumene. The emissions of the major compound and the sum of the total BVOCs are presented in Figure 16. The composition of the different mixtures (i.e., 3-hexen-1-ol, curcumene and other BVOCs) emitted for each temperature are presented in Figure 17.

Figure 16 shows that *Cistus albidus* emissions present two peaks, at 130 and 170 °C, and that between 130 and 150 °C emissions are decreasing. 3-hexen-1-ol emissions follow the same trend as total VOCs emissions and we can see that they represent an important part of the mixtures. The boiling temperature of 3-hexen-1-ol is around 160 °C and can therefore explain the decreasing above 170 °C by the thermal degradation of this compound coupled to thermal cracking reactions. Total BVOCs emissions are twice more important between 70 and 100 °C, the same increase is observed for 3-hexen-1-ol. Emissions are ten times more important between 70 °C (lowest emissions) and 170 °C (highest emissions). The emission peak at 130 °C can be explained by the fact that all the water contained inside the plant is evaporated at this temperature, thus increasing emission and transport phenomena of BVOCs.

Table 5. BVOCs emitted by *Cistus albidus*: name, retention time, chemical formula, chemical family and skeletal formula

BVOC	Retention time [min]	Chemical formula	Chemical family	Skeletal formula
3-hexen-1-ol	4.28	$C_6H_{12}O$	Alcohol	
β-elemene	15.71	$C_{15}H_{24}$	Sesquiterpene	
β-bourbolene	17.05	$C_{15}H_{24}$	Sesquiterpene	
α-muurolene	17.63	$C_{15}H_{24}$	Sesquiterpene	
β-caryophyllene	18.00	$C_{15}H_{24}$	Sesquiterpene	
aromendrene	19.05	$C_{15}H_{24}$	Sesquiterpene	
curcumene	19.57	$C_{15}H_{24}$	Sesquiterpene	
zingiberene	19.72	$C_{15}H_{24}$	Sesquiterpene	
α-selinene	19.94	$C_{15}H_{24}$	Sesquiterpene	

BVOC	Retention time [min]	Chemical formula	Chemical family	Skeletal formula
γ-cadinene	20.40	$C_{15}H_{24}$	Sesquiterpene	
β-bisabolene	20.55	$C_{15}H_{24}$	Sesquiterpene	
α-humulene	21.52	$C_{15}H_{24}$	Sesquiterpene	
Shyobunone	22.75	$C_{15}H_{24}O$	Sesquiterpenic ketone	
α-bisabolol	23.48	$C_{15}H_{26}O$	Sesquiterpenic alcohol	
α-eudesmol	28.04	$C_{15}H_{26}O$	Sesquiterpenic alcohol	

From Figure 17 it is seen that 3-hexen-1-ol is the major compound for *Cistus albidus* emissions. Indeed, it represents more than 68 % of the mixtures for all the temperatures except 165 °C, where its proportion is only 34 %. It can also be noted that the proportion of the second compound is very low, from 0.4 % at 70 °C to 11 % at 165 °C, and 3.7 % on average. The sum of other compounds represents between 8 and 55 % of the mixture depending on the temperature and is 24 % on average. Curcumene percentage is very low but the sum of sesquiterpenes is 23 % of the mixtures on average for all the studied temperatures. The other identified families (sesquiterpenic keton and alcohol) represent therefore a low percentage of the different mixtures, less than 5 % whatever the temperature is, except α-eudesmol that represents 20 % at 70 °C.

Cistus albidus is the only studied species that does not mainly emit a terpenoid compound. The major compound identified here is the same as the one identified by Peñuelas and Llusia (2001) for the natural emissions of non-terpenoid BVOCs by Cistus albidus. 3-hexen-1-ol is even the only compound identified by these authors. This is in agreement with the present study since this compound is much more important than the others at high temperatures. Other authors worked on the terpene emissions under natural conditions of *Cistus albidus*: Owen et al. (2001) and Ormeño et al. (2007). Owen et al. (2001) shown that this species emit small amount of α-pinene and limonene. Not identified here is any monoterpenes for this species, this difference can be due to the differences between temperatures or trapping techniques and they only identified traces of these compounds. On the contrary, Ormeño et al. (2007) identified zingiberene, curcumene and aromendrene as major sesquiterpenes. These authors also studied the influence of the nature of the soil (calcareous or siliceous). The composition does not change with soil characteristics but the amount of emitted compounds is higher for siliceous soils.

Pinus Pinea Emissions

Fourteen compounds were identified for *Pinus pinea* species emissions and the main component is limonene. Figure 18 illustrates the total BVOCs emissions as well as those of limonene at different temperatures.

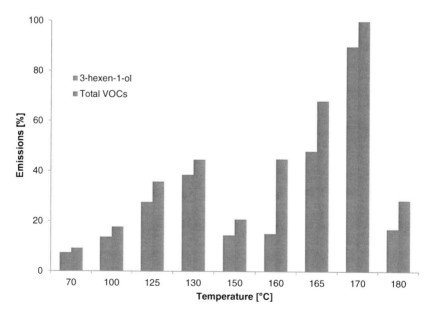

Figure 16. BVOCs relative emissions of *Cistus albidus* leaves at different temperatures.

Biogenic Volatile Organic Compounds Emissions of Heated ... 69

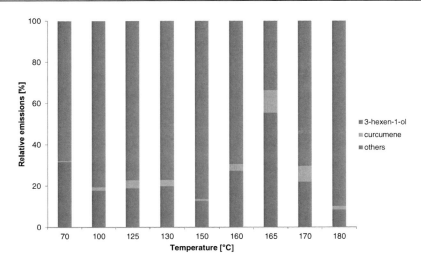

Figure 17. Composition of the BVOCs mixtures emitted by Cistus albidus leaves at different temperatures.

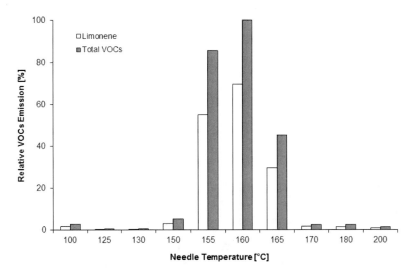

Figure 18. Limonene and total BVOCs emissions of heated Pinus pinea needles at different preheat temperatures.

This Figure exhibits that the BVOCs emissions increase with the needles temperature until 160 °C. The same trend is observed by Barboni et al. (2011) for other Mediterranean species such as needles of *Pinus nigra* and *Pinus pinaster* or shrubs of *Rosmarinus officinalis* as described above. Moreover, the higher value at 100 °C than the ones at 125 and 130 °C can be explained by the transport of the BVOCs by the water evaporation process. Knowing that the lowest boiling point of the identified compounds is the one of a-pinene and is about 150 °C (Zhao et al. 2011), these molecules would be in a liquid or in a vapour-liquid equilibrium state below this temperature. As a consequence, for the temperatures higher than this value, the BVOCs emission increases rapidly to a maximum at 160 °C. After this temperature, we can observe a decrease of the BVOCs amount that can be explained by the thermal degradation of the emitted terpenoids. The same BVOCs are obtained for all preheat temperatures and an example of the mixture composition at 160 °C (maximum BVOCs emissions) is given in Table 6. As we can see from this Table, the BVOCs mixture is characterized by high contents of limonene, a-pinene and b-caryophyllene. The total emissions of these constituents represent almost 83% of the total BVOCs emissions. Similar compounds are obtained by Macchioni et al. (2003) for the study of the BVOCs emissions of *Pinus pinea* needles, branches and female cones at ambient temperature. They have shown that limonene is the major constituent emitted by these three plant parts. Let us notice that the list proposed in this chapter is slightly different to the one of these authors because of differences in the studied temperature and in the season of needles samples collection. Indeed, they have worked with samples collected in winter (February) at ambient temperature. The emitted compounds can be classified into four chemical families according to their molecular formula. It is clear reading this Table that the BVOCs composition is characterized by a high monoterpenes ($C_{10}H_{16}$) percentage, higher than 84%. Sesquiterpene hydrocarbons ($C_{15}H_{24}$) are the second group and represent around 12% of the total mixture.

Comparison between the Five Vegetal Species

In this paragrah, the emissions of the five studied species are compared. First, it is noticed that *Cistus albidus* is the only species that does not emit mainly a terpenoid compound. Each vegetal species emit different chemical families and are presented in Figure 19 sorted by molecular formula. The repartition of the different group of molecules is presented at the temperatures where emissions are maximal.

Table 6. Relative emissions of different BVOCs from heated *Pinus pinea* needles at 160 °C

Compound	Molecular formula	Proportion in the VOCs mixture [%]
α-pinene	$C_{10}H_{16}$	8.23
β-pinene	$C_{10}H_{16}$	2.43
myrcene	$C_{10}H_{16}$	3.42
α-phellandrene	$C_{10}H_{16}$	0.13
limonene	$C_{10}H_{16}$	69.53
terpinolene	$C_{10}H_{16}$	0.51
o-cresol,6-tert-butyl	$C_{11}H_{26}O$	3.59
patchoulene	$C_{15}H_{24}$	2.72
β-caryophyllene	$C_{15}H_{24}$	4.86
α-selinene	$C_{15}H_{24}$	1.69
longifolene	$C_{15}H_{24}$	2.28
α-humulene	$C_{15}H_{24}$	0.24
germacrene D	$C_{15}H_{24}$	0.21
guaiol	$C_{15}H_{26}O$	0.17

Figure 19 highlights the fact that *Thymus vulgaris* and *Lavandula stœchas* emit a wider range of compound than *Rosmrinus officinalis*, *Cistus albidus* and *Pinus pinea*. Indeed, *Thymus vulgaris* and *Lavandula stœchas* emit six different groups of BVOCs, *Rosmarinus oficinalis* 5 and *Cistus albidus* and *Pinus pinea* only 4. *Cistus albidus* is also the only species that does not emit monoterpene ($C_{10}H_{16}$ or $C_{10}H_{14}$). It is also noticed that for all species major group represent more than 45 % of the mixtures, with some discrepancies between species; indeed, for *Rosmarinus officinalis*, $C_{10}H_{16}$ compounds represent 45 % of the total BVOCs, and the same group (also major) represent 85 % of the emissions of *Pinus pinea* at the maximal emission temperature. The weight of the second major group also depends on the species: it represents around 30 % for *Thymus vulgaris*, *Lavandula stœchas*, *Rosmarinus officinalis*, *Cistus albidus* for respectively $C_{10}H_{14}$, $C_{10}H_{18}O$, $C_{10}H_{18}O$ and $C_{15}H_{24}$ compounds but only 12 % for *Pinus pinea*, the second group being $C_{15}H_{24}$ compounds.

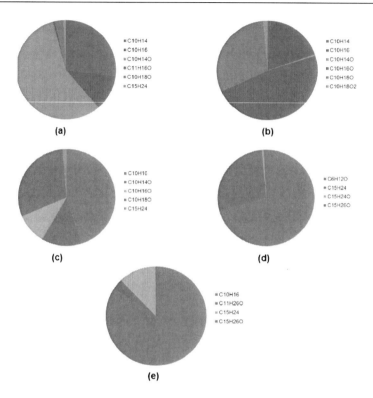

Figure 19. BVOCs (sorted by chemical formula) emitted by each vegetal species at the maximal emission temperature: (a) *Thymus vulgaris* at 180 °C, (b) *Lavandula stœchas* at 180 °C, (c) *Rosmarinus officinalis* at 165 °C, (d) *Cistus albidus* at 170 °C and (e) *Pinus pinea* at 160 °C.

In order to compare the amount of BVOCs emitted by each species (and since calibration was hard to perform), the emissions (in peak area divided by the sample mass) of the total BVOCs emitted by each vegetal species as functions of temperature is presented in Figure 20. It was pointed out above that the temperature of the emission peak depends on the boiling point of the major compound. This also appears in this figure and it is clearly seen that species that emit mainly an oxygenated monoterpenoid (i.e., *Thymus vulgaris* and *Lavandula stœchas*) reach their maximal emissions at the higher studied temperature (180 °C). Species that emit mainly a monoterpene (i.e., *Rosmarinus officinalis* and *Pinus pinea*) have their emission peak at a lower temperature. Oxygenated compounds have higher boiling point than the other compounds.

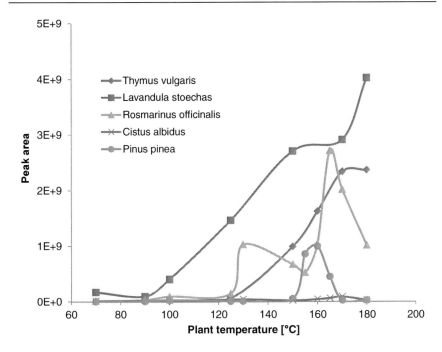

Figure 20. Emissions (in peak area) of the total BVOCs emitted by five vegetal species as functions of temperature: *Thymus vulgaris*, *Lavandula stœchas*, *Rosmarinus officinalis*, *Cistus albidus* and *Pinus pinea*.

Figure 20 also shows that *Lavandula stœchas* is the studied species that emit the highest amount of BVOCs, whatever the temperature is. *Rosmarinus officinalis* and *Thymus vulgaris* emit high amount of BVOCs. The maximal emissions of rosemary are higher than those of thyme, but thyme emits more than rosemary for more temperatures. It is also noticed that *Pinus pinea* emissions are negligible before 147 °C, and that even the peak is not very high: it is 4 times lower than the one of *Lavandula stœchas* and 2.8 times lower than the one of *Rosmarinus officinalis*. *Cistus albidus* emissions are negligible compared to the ones of the other studied species for all temperatures. The amount of BVOCs emitted by *Cistus albidus* at the emissions peak is 45 times lower than the amount emitted by *Lavandula stœchas* and 30 times lower than the one emitted by *Rosmarinus officinalis*.

CONCLUSION

This chapter deals with the study of the BVOCs emissions of five typical Mediterranean plant species at elevated temperature. These species, *Thymus vulgaris, Lavandula stœchas, Rosmarinus officinalis, Cistus albidus* and *Pinus pinea*, are known to be involved in forest fires. The study of the emissions has been carried out at middle and small scales. After determining the optimal heating time and heating rate for small scale experiments, emissions have been studied in the temperature range 70-180 °C using a flash pyrolysis apparatus. Emissions at middle scale have been investigated in an hermetic enclosure equipped with a radiant panel. Results show that high amount of terpenoid compounds are emitted, except for *Cistus albidus* emissions. The major compounds identified are thymol, 1-fenchone, a-pinene, 3-hexen-1-ol and limonene. These results provide a database that could help to improve the characterization of thermal degradation of vegetal fuels and to take into account the BVOCs combustion in physical forest fires models. The type of BVOCs emitted by the vegetal species tested and the amount produced indicate that under the right circumstances BVOCs from these vegetal species could contribute to the development of an accelerating forest fire. Some recent works (Peñuelas et al. 2013, Farré-Armengol et al. 2013) shown the possibility to detect BVOCs concentration at field scale by remote sensing methods that could permit to compare concentrations obtained during experiments with field ones. In a future work, we will investigate more deeply the gases accumulation with the determination of the dimensionless VOCs concentrations in a canyon using the wind tunnel measurements (Aubrun and Leitl 2004) and deducing these concentrations at real scale.

REFERENCES

Arif, AA; Shah, SM. Association between personal exposure to volatile organic compounds and asthma among US adult population, *International Archives of Occupational and Environmental Health*, vol. 80, 711-719, 2007.

Aubrun, S; Leitl, B. Development of an improved physical modelling of a forest area in a wind tunnel, *Atmospheric Environment*, vol. 38, 2797-2801, 2004.

Barboni, T; Cannac, M; Leoni, E; Chiaramonti, N. Emission of biogenic volatile organic compounds involved in eruptive fire, implications for the safety of firefighters, *International Journal of Wildland Fire*, vol. 20, 152-161, 2011.

Bernstein, JA; Alexis, N; Bacchus, H; Bernstein, IL; Fritz, P; Horner, E; Li, N; Mason, S; Nel, A; Oullette, J; Reijula, K; Reponen, T; Seltzer, J; Smith, A; Tarlo, SM. The health effects of non-industrial indoor air pollution, *Journal of Allergy and Clinical Immunology*, vol. 121, 585-591, 2008.

Billionnet, C; Gay, E; Kirchner, S; Leynaert, B; Annesi-Maesano, I. Quantitative assessments of indoor air pollution and respiratory health in a population-based sample of French dwellings, *Environmental Research*, vol. 111, 425-434, 2011.

Chetehouna, K; Barboni, T; Zarguili, I; Leoni, E; Simeoni, A; Fernandez-Pello, AC. Investigation on the emission of Volatile Organic Compounds from heated vegetation and their potential to cause an accelerating forest fire, *Combustion Science and Technology*, vol. 181, 1273-1288, 2009.

Cooke, KM; Hassoun, S; Saunders, SM; Pilling, MJ. Identification and quantification of volatile organic compounds found in a eucalyptus forest during FIELDVOC'94 in Portugal, *Chemosphere - Global Change Science*, vol.3, 249-257, 2001.

Courty, L; Chetehouna, K; Lemée, L; Mounaïm-Rousselle, C; Halter, F; Garo, JP. Pinus pinea emissions and combustion characteristics of limonene potentially involved in accelerating forest fires, *International Journal of Thermal Sciences*, vol. 57, 92-97, 2012.

Farré-Armengol, G; Filella, I; Llusia, J; Peñuelas, J. Floral volatile organic compounds: Between attraction and deterrence of visitors under global change, *Perspectives in Plant Ecology, Evolution and Systematics*, vol. 15, 56-67, 2013.

Gascoin, N; Fau, G; Gillard, P; Mangeot, A. Experimental flash pyrolysis of high density polyethylene under hybrid propulsion conditions, *Journal of Analytical and Applied Pyrolysis*, vol. 101, 45-52, 2013.

Giménez, A; Arqués, P; Arnaldos, J. Experimental study for characterizing the effectiveness of long-term forest fire retardants and the influence of fuel characteristics on the reduction of the rate of fire spread, *Forest Ecology and Management*, vol. 234 Supplement, pp. S236, 2006.

Granström, K. Emission of monoterpenes and VOCs during drying of sawdust in a spouted bed, *Forest Products Journal*, vol. 53, 48-56, 2003.

Greenberg, JP; Friedli, H; Guenther, AB; Hanson, D; Harley, P; Karl, T. Volatile organic emissions from the distillation and pyrolysis of vegetation, *Atmospheric Chemistry and Physics*, vol. 6, 81-91, 2006.

Harti, S; Cifredo, G; Gatica, JM; Vidal, H; Chafik, T. Physicochemical characterization and adsorptive properties of some Moroccan clay minerals extruded as lab-scale monoliths, *Applied Clay Science*, vol. 36, 287-296, 2007.

Hassiotis, CH; Dina, EI. The influence of aromatic plants on microbial biomass and respiration in a natural ecosystem, *Israel Journal of Ecology and Evolution*, vol. 56, 181-196, 2010.

International Handbook on Forest Fire Protection, Technical guide for the countries of the Mediterranean basin, Ministère de l'aménagement du territoire et de l'environnement (France), F.A.O. (Food and Agriculture Organization of the United Nations), http://www.fao.org/forestry/27221-06293a5348df37bc8b14e24472df64810.pdf.

Isidorov, VA; Vinogorova, VT; Rafałowski, K. HS-SPME analysis of volatile organic compounds of coniferous needle litter, *Atmospheric Environment*, vol. 37, 4645-4650, 2003.

Liodakis, S; Vorisis, D; Agiovlasitis, IP. Testing the retardancy effect of various inorganic chemicals on smoldering combustion of Pinus halepensis needles, *Thermochimica Acta*, vol. 444, 157–165, 2006.

Llusià, J; Peñuelas, J. Pinus halepensis and Quercus ilex terpene emission as affected by temperature and humidity, *Biologia Plantarum*, vol. 42, 317-320, 1999.

Macchioni, F; Cioni, PL; Flamini, G; Morelli, I; Maccioni, S; Ansaldi, M. Chemical composition of essential oils from needles, branches and cones of Pinus pinea, P. halepensis, P. pinaster and P. nigra from central Italy, *Flavour and Fragrance Journal*, vol. 18, 139-143, 2003.

Moukhtar, S; Couret, C; Rouil, L; Simon, V. Biogenic Volatile Organic Compounds (BVOCs) emissions from Abies alba in a French forest, *Science of The Total Environment*, vol. 354, 232-245, 2006.

Nezhadali, A; Akbarpour, M; Shirvan, BZ; Mousavi, M. Comparison of volatile organic compounds of Thymus vulgaris using hydrodistillation and headspace solid phase microextraction gas chromatography mass spectrometry, *Journal of the Chinese Chemical Society*, vol. 57, 40-43, 2010.

Nuñez-Regueira, L; Rodriguez-Anon, JA; Proupin, J; Mourino, B; Artiaga-Diaz, R. Energetic study of residual forest biomass using calorimetry and thermal analysis, *Journal of Thermal Analysis and Calorimetry*, vol. 80, 457-464, 2005.

Ormeño, E; Fernandez, C; Bousquet-Mélou, A; Greff, S; Morin, E; Robles, C; Vila, B; Bonin, G. Monoterpene and sesquiterpene emissions of three Mediterranean species through calcareous and siliceous soils in natural conditions, *Atmospheric Environment*, vol. 41, 629-639, 2007.

Ormeño, E; Céspedes, B; Sánchez, IA; Velasco-García, A; Moreno, JM; Fernandez, C; Baldy, V. The relationship between terpenes and flammability of leaf litter, *Forest Ecology and Management*, vol. 257, 471-482, 2009.

Owen, SM; Boissard, C; Hewitt, CN. Volatile organic compounds (VOCs) emitted from 40 Mediterranean plant species: VOC speciation and extrapolation to habitat scale, *Atmospheric Environment, vol.* 35, 5393-5409, 2001.

Owen, SM; Harley, P; Guenther, A; Hewitt, CN. Light dependency of VOC emissions from selected Mediterranean plant species, *Atmospheric Environment*, vol. 36, 3147-3159, 2002.

Owens, MK; Lin, CD; Taylor, CA; Whisenant, SG. Seasonal patterns of plant flammability and monoterpenoid content in Juniperus ashei, *Journal of Chemical Ecology*, vol. 24, 2115-2129, 1998.

Page, DW; van Leeuwen, JA; Spark, KM; Mulcahy, DE. Pyrolysis characterisation of plant, humus and soil extracts from Australian catchments, *Journal of Analytical and Applied Pyrolysis*, vol.65, 269-285, 2002.

Parra, R; Gasso, S; Baldasano, JM. Estimating the biogenic emissions of non-methane volatile organic compounds from the North Western Mediterranean vegetation of Catalonia, Spain, *Science of the Total Environment*, vol. 329, 241-259, 2004.

Pegoraro, E; Rey, A; Greenberg, J; Harley, P; Grace, J; Malhi, Y; Guenther, A. Effect of drought on isoprene emission rates from leaves of Quercus virginiana Mill., *Atmospheric Environment*, vol. 38, 6149-6156, Issue 36, 2004.

Peñuelas, J; Lluisia, J. Seasonal patterns of non-terpenoid C6-C10 VOC emission from seven Mediterranean woody species, *Chemosphere*, vol. 45, 237-244, 2001.

Peñuelas, J; Guenther, A; Rapparini, F; Llusia, J; Filella, I; Seco, R; Estiarte, M; Mejia-Chang, M; Ogaya, R; Ibañez, J; Sardans, J; Castaño, LM; Turnipseed, A; Duhl, T; Harley, P; Vila, J; Estavillo, JM; Menéndez, S; Facini, O; Baraldi, R; Geron, C; Mak, J; Patton, EG; Jiang, X; Greenberg,

J. Intensive measurements of gas, water, and energy exchange between vegetation and troposphere during the MONTES campaign in a vegetation gradient from short semi-desertic shrublands to tall wet temperate forests in the NW Mediterranean Basin, *Atmospheric Environment*, vol. 75, 348-364, 2013.

Raffalli, N; Picard, C; Giroud, F. Safety and awareness of people involved in forest fires suppression, Forest Fire Research & Wildland Fire Safety, Viegas (ed.), ISBN 90-77017-72-0, Rotterdam, 2002.

Rumchev, K; Spickett, J; Bulsara, M; Phillips, M; Stick, S. Association of domestic exposure to volatile organic compounds with asthma in young children, *Thorax*, vol. 59, 746-751, 2004.

Sezer, M; Bilgesü, AY; Karaduman, A. Flash pyrolysis of Silopi asphaltite in a free-fall reactor under vacuum, *Journal of Analytical and Applied Pyrolysis*, vol. 82, 89-95, 2008.

Simon, V; Luchetta, L; Torres, L. Estimating the emission of volatile organic compounds (VOC) from the French forest ecosystem, *Atmospheric Environment*, vol. 35 Supplement 1, S115-S126, 2001.

Viegas, DX; Simeoni, A. Eruptive behaviour of forest fires, *Fire Technology*, vol. 47, 303-320, 2011.

Vokou, D; Chalkos, D; Karamanlidou, G; Yiangou, M. Activation of soil resepiration and shift of the microbial population balance in soil as a response to Lavandula stœchas essential oil, *Journal of Chemical Ecology*, vol. 28, 755-768, 2002.

Yassaa, N; Williams, J. Analysis of enantiomeric and non-enantiomeric monoterpenes in plant emissions using portable dynamic air sampling/solid-phase microexctraction (PDAS-SPME) and chiral gas chromatography/mass spectrometry, *Atmospheric environment*, vol. 39, 4875-4884, 2005.

Zaitan, H; Bianchi, D; Achak, O; Chafik, T. A comparative study of the adsorption and desorption of o-xylene onto bentonite clay and alumina, *Journal of Hazardous Materials*, vol. 153, 852-859, 2008.

Zhao, FJ; Shu, LF; Wang, QH; Wang, MY; Tian, XR. Emissions of volatile organic compounds from heated needles and twigs of Pinus pumila, *Journal of Forestry Research*, vol. 22, 243-248, 2011.

In: Volatile Organic Compounds
Editor: Khaled Chetehouna

ISBN: 978-1-63117-862-7
© 2014 Nova Science Publishers, Inc.

Chapter 4

CONTRIBUTION OF BIOGENIC VOLATILE ORGANIC COMPOUNDS TO TROPOSPHERIC OZONE FORMATION IN THE PEARL RIVER DELTA REGION OF CHINA

K. Cheung and H. Guo[*]

Air Quality Studies, Department of Civil and Environment Engineering,
Hong Kong Polytechnic University

ABSTRACT

The Pearl River Delta (PRD) of China— encompasses 0.41% of China's land area but accounts for about 9% of China's national GDP— has been a rapidly developing economic region since 1980s. Like many other megacities in the world, it suffers from serious air pollution problems, particularly ozone (O_3) pollution. O_3, produced from a series of chemical reactions in the presence of nitrogen oxides (NO_X), volatile organic compounds (VOCs) and sunlight, is harmful to both human health and the environment. As a major precursor of tropospheric O_3, VOCs originate from both anthropogenic and biogenic sources. Biogenic VOCs (BVOCs), primarily composed of isoprene and monoterpenes, are emitted naturally in substantial quantities from certain types of terrestrial

---

[*] Corresponding author: Air Quality Studies, Department of Civil and Environmental Engineering, The Hong Kong Polytechnic University, Hung Hom, Kowloon, Hong Kong Email: ceguohai@polyu.edu.hk.

vegetation. The atmospheric reactivities of most BVOCs are higher than those of many anthropogenic VOCs (AVOCs), and thus, they are believed to play an important role in the formation of tropospheric O_3. In metropolitan cities, the potential of BVOCs to form O_3 is amplified by high concentrations of NO_x. Therefore, characterization of BVOCs is essential for understanding O_3 chemistry in urban areas, and for the regional air quality modelling. The following book chapter will summarize the findings from various air quality studies conducted in the PRD region, and provide a comprehensive review of the contribution of BVOCs to O_3 formation in a wide range of space and time. By thoroughly evaluating the implications of photochemical oxidation of BVOCs, more effective air quality regulations could be developed to control O_3 pollution.

Keywords: Biogenic volatile organic compound; Ozone formation; Pearl River Delta region; China

INTRODUCTION

Ozone (O_3) – formed via the chemical reactions of nitrogen oxides (NO_x) and volatile organic compounds (VOCs) under sunlight – is a major air pollutant in many urban centres. Numerous epidemiological and toxicological studies have associated measured O_3 levels with adverse health outcomes including decreased lung function, exacerbation of asthma, and premature deaths (Devalia et al., 1997; Gryparis et al., 2004; Haney et al., 1999; Lippmann, 1993; Yang et al., 2012). O_3 could also harm the environment by weakening sensitive vegetation and reducing agricultural yields. Since the 1980s, the rapid economic development in the Pearl River Delta region (PRD) in South China has led to a tremendous increase of energy consumption, resulting in regional air pollution. Hong Kong and the PRD are heavily impacted by photochemical O_3 and secondary organic aerosols (SOA) pollution. Controlling O_3 is challenging due to the nonlinear response of O_3 to its precursors. This relationship is specific to location; therefore an in-depth understanding of the sources and formation mechanisms of photochemical O_3 and its precursors, as well as their spatio-temporal variations, is needed to establish effective O_3 control strategies.

As a major precursor of tropospheric O_3, VOCs originate from both anthropogenic and natural sources. Over the last two decades, the relationship between O_3 and its precursors has been extensively studied in South China. O_3

formation is generally sensitive to VOC levels and composition, with toluene and isoprene (on a carbon basis) as the leading contributor to O_3 production in the urban and rural areas in Hong Kong, respectively (Cheng et al., 2010a; Zhang et al., 2007). In the PRD, isoprene was the most significant contributor to O_3 formation among the measured non-methane hydrocarbons (NMHCs), followed by m,p-xylene, ethylene and toluene in 2006 (Zheng et al., 2009b). Nonetheless, most of the studies were conducted in urban environment focusing on the impacts of the anthropogenic VOCs on O_3 formation. The overall influence of BVOC emissions on atmospheric photochemical processes remains to a large extent unknown.

It is estimated that about 80% of the global VOCs emissions are biogenic in origin (Warneck, 2000). Isoprene and monoterpene contribute 57% and 14%, respectively, to the total reactive VOCs emitted globally to the atmosphere (Guenther et al., 1995). The major emitted BVOC include isoprene (C_5H_8), monoterpenes ($C_{10}H_{16}$), sesquiterpenes ($C_{15}H_{24}$) and several oxygenated species (Ciccioli et al., 1993; Isidorov et al., 1985). Isoprene and monoterpenes are of particular interest because they play a critical role in the formation of tropospheric O_3 and SOA due to their high reactivity (Atkinson, 1997). The reactivity of BVOCs is often two to three times that of AVOCs (Carter and Atkinson, 1996). Once released into the atmosphere, these compounds react mainly with O_3, OH and NO_3 radicals, with the subsequent production or regeneration of O_3 and photochemical oxidants. Thus, even at low concentrations, BVOCs can have a significant impact on air quality. The impact of BVOCs on O_3 and secondary particle formation becomes more prominent in warmer seasons, when both photochemical reactivity and BVOC emissions are at a maximum. In areas with high vegetation coverage, BVOC could be the most important source category in driving O_3 production, and the O_3 produced could elevate the O_3 levels in downwind locations. Therefore, photochemical oxidation of BVOCs has important implications for local and regional air quality, particularly in areas having high vegetation coverage and climate conditions that favour BVOC emissions (Arneth et al., 2008; Chameides et al., 1992). Indeed, the importance of BVOC emissions in an O_3 inventory became apparent in regions when BVOC emission estimates were compared to the AVOC emission estimates (Chameides et al., 1988).

The PRD has high vegetable coverage because of the warm and humid subtropical climate, resulting in high BVOC emissions throughout the year. Due to complex topography and land use, mesoscale circulations could transport and redistribute urban air masses. In the last decade, some studies were conducted in Hong Kong and the PRD to understand BVOC contributions to local O_3 formation and its subsequent impact to downwind areas. This chapter provides a comprehensive assessment of the impact of BVOCs on O_3 chemistry in South China. By investigating the key chemical processes in ozone-precursor relationships in polluted urban environments, this study will help in formulating an effective O_3 control strategy.

METHODS

A search of the Web of Science (ISI) database was conducted on February 1, 2014. The search algorithm started with a combination search of (1) topic of interest and (2) study area of interest. A Title/Keyword/Abstract search using the terms *"BVOC,"* *"Isoprene,"* or *"Monoterpene"* was used to capture the topic of interest, while the terms *"Hong Kong,"* or *"Pearl River Delta,"* were used to specify the study area. For the resulted articles, their abstracts and full texts were obtained to determine each publication's relevance to our review. Articles were considered irrelevant if BVOC emissions/levels were not reported, or if the study area excluded the PRD/Hong Kong. 18 relevant peer-reviewed journal articles were obtained and included in this review.

To evaluate the impact of BVOC emissions on O_3 formation, it is necessary to characterize the abundance of BVOCs. BVOC abundance varied spatially and temporally depending on vegetation types, temperature, light flux, and other meteorological parameters. There are 3 distinct methods to estimate BVOC abundance: (1) atmospheric measurement, (2) emission inventory and (3) remote sensing. The 18 journal articles are segregated into the 3 categories and discussed below.

ATMOSPHERIC MEASUREMENT

In the last few decades, canister sampling has been used extensively to characterize ambient VOC levels. The quantification of isoprene began as early as 2000 in Hong Kong. So and Wang (2004) studied the spatio-temporal variation, source–receptor relationships, and photochemical reactivity of $C_3 - C_{12}$ NMHCs from different areas in Hong Kong based on VOC canister samples collected from November 2000 to October 2001. Four sampling locations, namely a rural coastal site (Hok Tsui, HT), an urban industrial site (Central/Western, CW), an urban residential site (Tsuen Wan, TW) and a roadside site (Mong Kok, MK), were used. The annual average concentration of isoprene at the rural, residential, industrial, and roadside sites was 0.46 ppbv, 0.30 ppbv, 0.43 ppbv and 0.65 ppbv, respectively. Biogenic emission was the major source of isoprene at the rural site. The highest monthly average occurred in August, when the temperature was higher thereby facilitating the biogenic emissions of isoprene. Although higher isoprene levels were observed in August and September of 2001 at all sites, high levels of isoprene were also recorded in springtime. At the roadside sites, the highest monthly average of isoprene was observed in March, when the concentrations of vehicular markers such as benzene were also higher. Thus, vehicular emission could make an important contribution to isoprene levels in urban areas of Hong Kong. The ozone formation potential (OFP) was also evaluated at the four sites. Isoprene had the highest OH-reactivity and OFP at the rural site, while toluene was the most important contributor at the roadside site. Nonetheless, isoprene was still a key contributor to OFP at the urban sites and roadside site, ranking fifth and sixth, respectively, among the measured NMHCs species (So and Wang, 2004).

From September 2002 to August 2003, 248 ambient VOC samples were collected at four Hong Kong locations ranging from urban (Central/Western), sub-urban (Yuen Long and Tung Chung) to rural (Tap Mun) areas (Guo et al., 2007). Tap Mun exhibited the highest isoprene levels (average = 334 pptv, 95% CI = 81 pptv) due to the presence of significant biogenic sources of isoprene (i.e., vegetation). At this site, the elevated levels of BVOC made significant contributions to O_3 formation as compared to anthropogenic VOCs. The average isoprene levels at the urban and suburban sites ranged from 148-192 pptv. Seasonal variations were similar at all sites. Isoprene concentrations were the lowest in the winter, and began to increase in late spring. Higher monthly concentrations occurred from June to September. The seasonal variations were more pronounced at Tap Mun, with monthly average mixing ratio up to approximately 750 pptv during summertime. A similar sampling campaign was conducted from September 2006 to February 2008. Lau et al. (2010) studied the major sources of VOC at the four stations based on data from these two sampling periods (i.e., August 2002 to August 2003 and September 2006 to August 2007) using positive matrix factorization (PMF). The PMF-estimated biogenic source contributions to total VOCs were 2-5.5% from September 2002 to August 2003, and 1.3-4.4% from September 2006 to August 2007 (Lau et al., 2010). As expected, biogenic emissions showed the largest impact on the rural site (Tap Mun), followed by the sub-urban site at Tung Chung. Biogenic emissions were higher in summer and lower in winter at all four stations during both sampling campaigns, consistent with the results from the previous study (So and Wang, 2004).

On the other hand, another study conducted in downtown Guangzhou and rural Xinken revealed that biogenic emission was not a dominant contributor to the high O_3 levels (Wang et al., 2008). The sampling was conducted from October 16 to November 5, 2004, with an average temperature of 25°C during the sampling campaign. Weak midday isoprene peaks were observed at both sites, and correlation analysis suggested that isoprene in downtown Guangzhou likely originated from fresh traffic emissions. Although significant levels of isoprene were recorded in Guangzhou (average = 0.49 ppbv) and Xinken (average = 0.46 ppbv), the lack of strong midday emissions suggested that BVOC might not impact O_3 formation significantly during the sampling period.

Tang et al. (2007) investigated the characteristics and diurnal variations of NMHCs at urban (Guangzhou, GZ) and suburban (Panyu, PY) Guangzhou, a rural site in the PRD (Dinghu mountain, DM) and a remote site in South China (Jianfeng mountain, JM) in April of 2005. The average isoprene levels were 0.27 ± 0.14 ppbv, 0.18 ± 0.10 ppbv, 0.12 ± 0.80 ppbv, 0.48 ± 0.47 ppbv at GZ, PY, DM and JM, respectively. Isoprene, generated from local biogenic emission, was the main contributor to the total hydrocarbon at the remote site (JM). The diurnal variations of isoprene were similar at the two mountain sites. Isoprene levels started to increase in the morning and reached peak levels at around 2 p.m., and then decreased steadily until late evening. The two sites in Guangzhou, on the other hand, experienced negligible diurnal variations, and the investigators attributed the higher levels of isoprene at the GZ site to vehicular emission of isoprene in the Guangzhou urban area (Tang et al., 2007).

In July 2006, another study was conducted to provide a wider spatial coverage of BVOC abundance in the PRD region. Measurements of VOC concentrations and compositions were performed at six sampling locations (Yuan et al., 2012). Two sites were established in the urban area of Guangzhou (Baiyunshan, BYS) and Foshan (Huijingcheng, HJC). The other four regional sites were more rural in nature, located at Wanqingsha (WQS), Zhuhai Tangjia (TJ), Conghua Tianhu (TH), and Huizhou Jinguowan (JGW). The average isoprene levels were 1.14 ± 0.54, 1.47 ± 0.65, 0.79 ± 0.43, 2.86 ± 1.07, 1.66 ± 0.40 and 1.08 ± 0.42 ppb for BYS, HJC, WQS, TJ, TH and JGW respectively. These levels were much higher than those reported in other studies (Tang et al., 2007; Wang et al., 2008), mainly because the VOC samples were collected in summertime, when the temperature was much higher than those in other

Contribution of Biogenic Volatile Organic Compounds ...   85

sampling campaigns. TJ, a tourist city with high vegetable coverage and little industrial activity, recorded the highest isoprene concentration. The investigators calculated the OFP based on the products of ambient concentrations of the species and their maximum incremental reactivity (MIR). The OFP was the highest at TJ (~40 ppb), followed by TH (~24 ppb) and HJC (~21 ppb). The average OH loss rate (R_{OH}), calculated as the products of ambient concentration of the species and the corresponding rate constant of the species reacting with OH radical, was used to evaluate the reactivity of VOCs from these ambient measurements. Anthropogenic alkenes and isoprene were the two major contributors to R_{OH}, especially at HJC, TH and TJ. In particular, isoprene contributed 60.3% of the OH loss rate at TJ (Yuan et al., 2012). A similar campaign was conducted from October to November, 2008 at 2 sites: TH and Jiangmen Kaiping (JM), and they were chosen as upwind and downwind sites of urban Guangzhou, respectively. The average isoprene levels were 0.20 ± 0.15 ppb at JM and 0.38 ± 0.55 ppb at TH. Note that R_{OH} and OFP were much lower in the 2008 than 2006, likely because the temperature in fall 2008 was much lower than that in July 2006. OFP at the TH site dropped from ~ 24 ppb in 2006 to less than 10 ppb in 2008, highlighting the importance of temperature and light in the contribution of isoprene (Yuan et al., 2012).

To evaluate the spatio-temporal characteristics of VOCs in South China, 198 air samples were collected simultaneously at a background at Wanqingsha (WQS), Guangzhou of inland PRD and Tung Chung (TC) in Hong Kong between October 25 and December 1, 2007. At TC, the average levels for isoprene, α-pinene and β-pinene were 270 ± 260 pptv, 140 ± 400 pptv and 60 ± 310 pptv respectively. The corresponding concentrations at the background WQS site were 140 ± 170, 110 ± 270 and 20 ± 20 pptv respectively (Cheng et al., 2010b). At the WQS site, a photochemical pollution episode was observed between November 12 and 17, 2007. A photochemical trajectory model (PTM), coupled with the Master Chemical Mechanism (MCM) was used to assess photochemical O_3 formation in the PRD region by calculating the photochemical ozone creation potential (POCP) indices (Cheng et al., 2010a). The investigators revealed that mobile sources were the largest contributor to regional O_3 formation (40%), followed by biogenic sources (29%), and that regional scale O_3 formation in the PRD region was mainly attributed to a relatively small number of VOC species, namely isoprene, ethene, m-xylene, and toluene. While the PTM revealed isoprene as the most important VOC species to generate O_3 in the PRD, the results generated by an Observation-Based Model (OBM) found that isoprene had a negligible effect on O_3

formation at the WQS site using the same dataset (Cheng et al., 2010c) . The investigators attributed the difference to the fact that POCP-weighted values in the PTM incorporated the contributions of VOCs to O_3 formation at a regional scale, while the OBM results were more site-specific. In the same period, the impact of BVOC was more pronounced in Hong Kong (Tung Chung site). Isoprene was the second most contributor to O_3 formation at Tung Chung, Hong Kong, with a relative incremental reactivity (RIR) value of ~0.2.

From September to November, 2010, VOC samples were collected at an urban site (Tsuen Wan) in Hong Kong (Cheung et al., 2014). Average isoprene levels were (252 ± 204 pptv) during the sampling period, with higher levels observed in the beginning of the sampling campaign (i.e., September 28 to October 24) when the temperature was higher (average daily temperatures of 30.6 ± 1.6 °C) compared to the latter part of the sampling period (i.e., October 27 to November 21). A typical bell-shaped distribution, with peaks between 11 a.m. and 3 p.m., was observed on most days (Cheung et al., 2014). Isoprene photochemical ages ranged between 12 and 42 minutes, suggesting that the measured isoprene was mostly generated from nearby country parks, and that it was not significantly impacted by regional air mass transport from the PRD region of China. In the same period, lower average isoprene mixing ratio was recorded (109 pptv) at a mountain site (elevation = 640 m) about 7 km northeast of the Tsuen Wan site (Guo et al., 2012). The isoprene photochemical ages were between 10 and 64 minutes at the mountain site. Both studies suggested that the isoprene measured in Hong Kong was not influenced by the PRD. Nonetheless, it is possible that the isoprene transported from mainland China may be lost in photochemical reactions as it moved to Hong Kong due to the high reactivity of isoprene. Thus, although the isoprene generated from mainland China did not participate in the local photochemistry in Hong Kong, it could have brought along elevated levels of its photochemical products and O_3. Indeed, the O_3 levels measured at the mountain site was much higher than those measured at the foot of the mountain, and the investigators revealed that the higher O_3 levels at the mountain were somewhat influenced by regional transport from the PRD. Based on a multi-day O_3 episode derived from this sampling campaign, biogenic emission contributed to 7 ± 1% to the measured VOC (Lam et al., 2013). In addition, Ling and Guo identified seven major sources of VOC using the PMF model, and their contributions to O_3 formation were quantified using the OBM. The relative incremental reactivity (RIR)-weighted values (taking into account both RIRs and the emission amount of each VOC source category) was ~6% for biogenic emissions, and isoprene's ranked the fifth

Contribution of Biogenic Volatile Organic Compounds ... 87

among 41 major VOC species in contributing to O_3 formation (Ling and Guo, 2014).

The quantification of VOC levels in atmospheric measurement studies is mostly conducted using canister sampling followed by gas chromatography analysis. The uncertainty of the sample collection is strongly affected by the procedure of sample filling, the nature of the inner surfaces of the canisters and the processes used for the cleaning of these surfaces. It could be minimized by better designs, following standard protocols and good quality assurance/quality control. The uncertainties associated with the offline chromatography analysis are rather small for the measured VOCs (Colman et al., 2001). Nonetheless, the 7-day recovery rate was found to be lower in isoprene (75 ± 8%) than other VOCs such as alkanes (~92%), aromatics (~87%) and alkenes (~85%), due to the high reactivity of isoprene (Hsieh et al., 2003). On the other hand, the reactivity rate of isoprene with OH radical is high (1.1×10^{-10} cm^3 molecues^{-1} sec^{-1}), around 17 times higher than toluene and 87 times higher than propane. Since reactive compounds degraded along the transport from the source to the receptor, the effect of photochemical reaction loss of isoprene could be significant in source apportionment studies; thereby resulting in an underestimation of BVOC's contribution to total VOC/NMHC/O_3 formation due to the depletion from source to receptor.

EMISSION INVENTORY

In addition to ambient measurements, the use of an emission inventory coupled with air quality modelling is another tool to characterize BVOC emissions. An emission inventory accounts for the amount of pollutants discharged into the atmosphere originating from all source categories in a certain geographical area within a specified time span, where air quality models use such information to stimulate emission fluxes. Wei at al. studied the impacts of BVOCs on O_3 formation in a tropical cyclone-related episode using a regional chemical transport model (Community Multi-scale Air Quality Model, CMAQ) coupled with the non-hydrostatic meteorological model MM5 (Wei et al., 2007). An anthropogenic emission inventory developed for a variety of species for Asia in 2000 by Streets et al. (2003) was adopted. To estimate BVOCs, the Biogenic Emissions Inventory System version 3.09 (BEIS 3.09) algorithms in Sparse Matrix Operator Kernel Emissions (SMOKE) model system were used. The biogenic emissions of isoprene, terpene and other reactive VOCs were 8,500, 3,400 and 11,300 ton per day, respectively, in this 3-day episode event from September 14 to 16, 2004. Isoprene contributed to 36.4% of the total BVOCs. The simulations for their impacts on O_3 were done with and without the incorporation of biogenic emissions, while the anthropogenic emissions and meteorological conditions remained the same in the two test runs. It was found that the PRD was the area that had the highest response to biogenic emissions among other areas in South China. More O_3 was produced when biogenic emissions were included in the model, in comparison to that without BVOCs. Overall, the daily maximum difference in the modelled O_3 with and without BVOCs ranged from 1.2–4 ppb in South China and 4.6–15 ppb in the PRD region. The maximum impact of biogenic emissions on O_3 formation occurred in the afternoon, with the difference ranging from 21 to 27 ppb when maximum O_3 levels were in the order of 150–200 ppb. The investigators concluded that the PRD region might be the origin of biogenic impact, and that the additional O_3 due to biogenic impact could influence other areas depending on the prevalent meteorological conditions.

A more comprehensive assessment of BVOC emission in the PRD was conducted by Zheng et al. (2010). The investigators created an emission inventory for the year of 2006 using the Global Biosphere Emissions and Interactions System model, hourly BVOC emissions, newly available land cover database, observed meteorological data, and recent measurements of emission rates for tree species in China. The results show that the total annual BVOC emission was 2.2×10^{11} g C. Isoprene contributed about 25% (6.4×10^{10} g C), while the contributions of monoterpenes and other VOCs (OVOC) were about 34% and 41%, respectively. Peak emission was observed in July and the emissions were mainly distributed over the outlying areas of the PRD region, where the economy and land use were less developed. Minimum emission fluxes were observed in metropolitan urban areas (central Guangzhou-Foshan-Dongguan-Shenzhen), and their difference with the maximum emission fluxes in the outlying areas was about two orders of magnitude. Using the emission inventory developed by Zheng et al. for the PRD in 2006, biogenic emission (295.8 kt / yr) was the biggest source category of total VOC emissions, followed by gasoline vehicle (227.3 kt / yr) and motorcycle (212.6 kt / yr) (Zheng et al., 2009a). The source contributions of BVOC varied among individual cities, ranging from 8% at Foshan to 66% at Huizhou. Toluene was the biggest VOC contributor among the NMHCs, accounting for 9% of total VOC contribution, followed by isoprene, which accounted for 6.5% of VOC contribution. However, isoprene ranked the top in terms of ozone formation potentials (OFPs), accounting for 15.4%.

A BVOC emission inventory specific to Hong Kong was first assessed by Tsui et al. (2009) using 13 local tree species. Tree distribution of country parks was estimated based on field survey. From August 2004 to July 2005, the annual BVOC emission was 8.6 $\times 10^9$ g C in Hong Kong, calculated using plant emission data obtained from measurements and the literature, tree distribution estimation data, land use information, and meteorological data. Isoprene accounted for 30% (2.6 $\times 10^9$ g C) of the total BVOC emissions. Monoterpenes and other VOCs contributed about 40% and 30% respectively. Estimated BVOC emissions were the highest in summer, reaching 3.3 $\times 10^9$ g C, compared to about 0.96 $\times 10^9$ g C in winter. Diurnal variations were also predicted by the model. Isoprene, depending on both temperature and light, experienced a bell-shaped distribution with higher emission intensity from 11 a.m. to 3 p.m. Total monoterpenes and other VOCs, on the other hand, were estimated to be temperature dependent but light independent. Thus, they had a more steady emission throughout the day. Since BVOC emissions were mostly contributed by country parks where vegetation densities were higher than most of the non-country park areas, the spatial distributions of total BVOCs, isoprene, total monoterpenes and other VOCs were higher in country park areas including Lantau South and Lantau North Country Parks, Tai Lam Country Park, Lam Tsuen Country Park, and Sai Kung West Country Park.

The major components of an emission inventory assessment are land use distributions, emission factors, other input parameters such as meteorological parameters, and the algorithms relating emissions to weather (Tsui et al., 2009). Each of these components can make a significant contribution to the total uncertainty associated with emissions estimates (Guenther et al., 2006; Hanna et al., 2005). Since emission inventories often have different origins, objectives, as well as spatial and temporal resolutions, uncertainties from emission input are regarded as the largest source of uncertainties (Placet et al., 2000; Russell and Dennis, 2000). Emission uncertainties are often quantified using propagation of input uncertainties, encoding of expert elicitation, and inverse air modeling. When vegetation type and composition are characterized, and emission factor measurements are available for the dominant species in a region, emission estimates are generally within 50%. However, uncertainties could be a factor of 5 or more when these data are not available (Guenther et al., 2000; Guenther et al., 2006). The 95% confidence ranges on the calculated uncertainties for isoprene covered approximately an order of magnitude, and were much larger than those for monoterpene and OVOCs (<±20%) (Hanna et al., 2005). In sub-tropical Hong Kong, vegetation type and species are extremely diverse with up to 300 different tree species in one hectare; yet, less than 10% of the tropical species have been measured (Nichol and Wong, 2011; Tsui et al., 2009). For the abovementioned studies, the plant species representing plant-functional types / vegetation class were sampled from a finite number of field surveys. Emission factors used were sometimes determined by literature or taxonomy, which were often not specific to local values in Hong Kong. Given the high theoretical number of species in Hong Kong, a large gap between the theoretical number of plant species and the surveyed number of species is expected (Leung et al., 2010). The quantification of uncertainties in model parameters is typically done with statistical methods, such as bootstrap simulation. Due to the extensive variability in emissions among vegetation species, and the lack of detailed spatial characteristics on plant biomass, BVOC emission estimates driven by emission inventory and modelling are subject to large uncertainties.

REMOTE SENSING

In the last few years, the technique of remote sensing has been applied to map BVOC emissions. In 2009, Leung et al. (2010) developed an improved land cover and emission factor database to estimate Hong Kong's BVOC emissions using the Model of Emissions of Gases and Aerosols from Nature (MEGAN, a global BVOC emission model developed by Guenther et al. (2006)). The emission rates of isoprene and monoterpene of 13 local tree species were determined by laboratory measurements (Tsui et al., 2009). To supplement the existing tree survey data, 61 additional field surveys of plant species composition were conducted. A habitat map based on satellite images (10-m resolution) was used to provide high-resolution land cover data. The BVOC emissions from Hong Kong were calculated for 12 consecutive years from 1995 to 2006. Weather fluctuations imposed a small year-to-year emission variability over the 12-year period. Yet, an increasing trend in the annual variation can be observed due to an increase in forest land cover, suggesting the importance of accurate land cover inputs for biogenic emission models. Seasonal variations, with higher emissions in summer and lower emissions in winter, were observed. In 2006, the annual BVOC emission was 9.82×10^9 g C, with the highest contribution in summer (44%), followed by autumn (26%). Isoprene was the highest contributor, accounting for 72% of the total BVOC emission. Average diurnal variations were similar in summer and in winter, with emissions mainly in day time. High emission of isoprene occurred for regions dominated by broadleaf trees, including north and northeast New Territories, and part of the Lantau Island. On the other hand, the spatial variation of monoterpenes was different from that of isoprene, with higher emissions observed in areas with mangroves (Leung et al., 2010).

Nicole and Wong (2011) applied the global BVOC model, developed by Guenther et al. (1995), at a spatially detailed level to Hong Kong's landscape using high resolution remote sensing and ground data. Detailed maps of temperature, foliar density, and photosynthetically active radiation (PAR) were obtained by field work combined with ASTER, SPOT and Landsat satellite images. The spatial resolution was 10 meters with a half-hourly temporal resolution. BVOC emission rates were assigned to ecosystem types but not individual species due to the diverse vegetation types in Hong Kong. The modelled isoprene emissions were higher in summer than in winter, with higher emissions in central and eastern Hong Kong, where the mountains and country parks were located. The investigators validated the model using field data collected in 2007. The model-derived BVOC flux distributions showed reasonable consistency with the field observations, suggesting the application of remote sensing to BVOC mapping is promising.

More recently, Wong et al. (2013) modelled hourly isoprene emissions in Hong Kong using a geographic information system (GIS) and remote sensing database. Input parameters of temperature, foliar density, and PAR were based on satellite image data resampled to a 100 m resolution. A land use / land cover map at 10 m resolution was used to estimate the distribution map of vegetation types using nine classes of land use/land cover, namely industrial, residential, agricultural, mixed forest, grassland, shrubland and shrubby grassland and others. The isoprene emission capacity derived from Guenther et al. (1995) and Tsui et al. (2009) was assigned to the five vegetation classes. To calculate isoprene emission rates, a program was written in ESRI® ArcGIS^{TM}. First, a set of mesh polygons at 100 m resolution was created, and the land cover / land use data was used to assign the 9 different land types to the grids. Hourly maps of isoprene emission rates, calculated using isoprene emission capacity, foliar density, PAR and temperature factor, were then derived for each polygon. The modelled isoprene flux (mg C m^{-2} h^{-1}) showed significant spatial and temporal variation, with higher fluxes in central and eastern Hong Kong, as well as southern Hong Kong Island and part of the Lantau Island. The difference between the fluxes of summer (July 10, 2007) and winter (January 18, 2008) could be up to a factor of 10. The yearly isoprene emission from February 2007 to January 2008 was 6.8 $\times 10^8$ g C per year. The modelled isoprene emissions were validated using ambient isoprene concentrations from field measurements, where high correlations were found (R^2= 0.63 in summer and 0.37 in winter).

The uncertainties derived from the remote sensing studies are similar to those by the use of an emission inventory, because they use similar parameters such as standard emission rates, levels of foliar density, weather data, etc. to obtain BVOC emission estimates. In studies with the use of remote sensing data, some input parameters are estimated using satellite images where no information is available to judge their uncertainties. The different land use / land cover data used in each study could also drive significant difference in the emission estimates. For example, Leung et al. used an improved land cover and emission factor database to estimate BVOC emission in Hong Kong (Leung et al., 2010). During the period from August 2004 to July 2005, Leung et al. showed higher isoprene emission (+104%) and lower monoterpenes emissions (-65%) compared to the results obtained earlier by Tsui et al. (2009), and the investigators attributed the large differences to the change in land cover data (Leung et al., 2010). The use of high resolution satellite images could potentially improve the estimation of land cover / vegetation type and distribution in subtropical Hong Kong where high species and ecosystem diversity were found. The resulting higher spatial resolution data would also allow cross-validation with field measurements, thereby improving the reliability of the resulting estimates. While the abovementioned studies mapped isoprene emission in Hong Kong, they did not quantify its impacts on O_3 chemistry. Nonetheless, these studies demonstrated a simple and relatively inexpensive way to estimate isoprene emission at a detailed level in Hong Kong. The results could be used to further our understanding on the impacts of BVOC emission on the generation of tropospheric O_3.

BVOC Chemistry

Once released into the atmosphere, BVOCs undergo atmospheric processing and evolve with the changing ambient conditions. Key parameters that influence this atmospheric evolution include levels of reactants, and meteorological conditions such as temperature, wind speed and direction, mixing height, etc. In daytime, reaction with OH radicals is the main removal pathway of isoprene. The night-time oxidation, on the other hand, is primarily accomplished by O_3 and NO_3 radicals (Apel et al., 2002; Brown et al., 2009). The reaction of isoprene oxidation products and NO results in NO_2, which rapidly photolyzes to yield O_3 (Biesenthal et al., 1997). In high NO_x

Contribution of Biogenic Volatile Organic Compounds ... 93

environments including the PRD and Hong Kong, formaldehyde, methyl vinyl ketone (MVK) and methacrolein (MACR) are the dominant oxidation products of isoprene, accounting for more than 50% of the carbon yield (Carter and Atkinson, 1996; Miyoshi et al., 1994; Zhao et al., 2004). These compounds react mainly with O_3, OH and NO_3 radicals, with the subsequent production or regeneration of O_3 and photochemical oxidants.

Given the sensitivity of isoprene emission to ambient light and temperature, its impact on O_3 and SOA formation could have a strong seasonal dependence. Indeed, the year-long studies showed higher BVOC / isoprene emission in summer compared to other seasons at urban and rural locations in Hong Kong (Guo et al., 2007; So and Wang, 2004). In intensive studies, much lower isoprene levels / contributions were observed when the samples were collected in the spring (Tang et al., 2007) or the autumn (Cheng et al., 2010b; Guo et al., 2012; Wang et al., 2008). In particular, a study conducted in July of 2006 revealed much higher isoprene levels, with 5 out of 6 sampling sites recorded average concentrations above 1 ppb in the PRD (Yuan et al., 2012). At one of the sites, sampling was repeated in October and November of 2008. Isoprene levels were much lower in fall (0.38 ± 0.55 ppb) than in summer (1.66 ± 0.40 ppb), highlighting the high seasonal dependence of isoprene. The seasonal variations observed in atmospheric measurement studies are consistent with the results from emission modelling. In the PRD of China, BVOC emissions were ~44 kt in July and ~10 kt in January (Zheng et al., 2010). Thus, isoprene abundance was around 4 times higher in summer than in winter. Note that at roadside sites, or in locations where the impacts of vehicular emissions were strong, the monthly variation of isoprene might not be dependent on seasons, but rather on the levels of vehicular emissions (So and Wang, 2004; Wang et al., 2008). In general, the impacts of BVOC emissions could be much higher in summer, especially when temperature is > 30°C in areas with high vegetation cover.

Many of the abovementioned studies showed that isoprene was the most important contributor to O_3 formation at rural sites (So and Wang, 2004; Yuan et al., 2012), and isoprene remained as a key contributor at urban sites based on OFP calculations in atmospheric measurement studies (So and Wang, 2004; Yuan et al., 2012). Nonetheless, assessments conducted using air quality models showed the significant role of spatial resolution / scale in quantifying the impacts of isoprene / BVOC. Air quality models, such as PTM, MCM and OBM, were used to quantify the impacts of BVOCs. In a tropical cyclone-related O_3 episode in the PRD region, a regional chemical transport model (CMAQ) estimated that the maximum impact of biogenic emission on O_3

formation occurred at 2 p.m., with the difference ranging from 21 to 27 ppb when maximum O_3 levels were in the order of 150–200 ppb in South China (Wei et al., 2007). Using the PTM, biogenic source contributed ~29% to O_3 formation in the PRD on a regional scale during a photochemical pollution episode in the fall of 2007. Meanwhile, the impacts of biogenic hydrocarbons were less significant when assessed using the OBM, which focused on the impacts at a specific site. The contribution of biogenic emission was around 6% estimated by the OBM using the data collected in the fall of 2010 in Hong Kong. Overall, the impact of BVOC on O_3 appears to be higher on a regional scale.

FUTURE CHALLENGES

Currently, the scientific community has focused on the two major BVOC species, namely isoprene and monoterpene, which constitute a major fraction of BVOCs. While other BVOCs such as alkanes, other alkenes, carbonyls, alcohols, esters, and acids also contribute significantly to BVOC emission (Atkinson and Arey, 2003), limited information is available on their source characterization. In the PRD, the vegetation is highly diverse with about 3200 native and introduced vegetation species and varieties (Hong Kong Herbarium, 2004). Yet, the BVOC species input into photochemical models are represented by only a few species such as isoprene and pinene, whereas more than 70 BVOC species are known to be emitted from vegetation (Isidorov et al., 1985). Therefore, to improve our understanding on the impact of BVOCs, more species of BVOCs need to be identified and quantified.

In addition, levels and diversity of BVOC emission vary widely among different plant species. For example, isoprene is mainly emitted from deciduous (hardwood/ broad leaf) trees including oak and willows, while monoterpenes predominately arise from coniferous (softwood) trees such as pines, cedars and firs (Andreaniaksoyoglu and Keller, 1995). Emission factors from different vegetation have been extensively studied in the forest areas of North America (Helmig et al., 2007; Helmig et al., 2006; Isebrands et al., 1999) and Europe (Janson et al., 1999; Pokorska et al., 2012). Although tropical and sub-tropical regions are the dominant source of BVOCs, measurements / laboratory studies are limited and largely confined to South Africa (Otter et al., 2002; Saxton et al., 2007) in these regions. Given the diversity and specificity of vegetation in Southeast Asia, it is essential to better characterize the BVOC emission of local plants. Emission factors of

Contribution of Biogenic Volatile Organic Compounds ... 95

vegetation types specific to the PRD and Hong Kong need to be established and cross-validated with measurement studies.

Finally, the studies included in this review mainly focused on the impacts of BVOCs on O_3 formation in the PRD region. In the last few years, the contribution of primary and secondary source to particle-bound organic carbon has been studied in the PRD and Hong Kong (Duan et al., 2007; Hu et al., 2010; Huang et al., 2012). In urban locations with strong primary emissions such as Guangzhou, SOA was a minor component of $PM_{2.5}$ (Duan et al., 2007). On the contrary, SOA accounted for a significant fraction of the measured organic carbon (OC) at the urban and sub-urban sites in Hong Kong, with about half of the measured OC accounted by SOA on days influenced by regional sources (Hu et al., 2010). Despite the significance of secondary source contribution, limited studies have examined the role of BVOCs in SOA formation in the PRD region (Ding et al., 2012; Hu et al., 2008; Jiang et al., 2012; Wang et al., 2013). With new techniques such as remote sensing to improve the land cover data and the development of emission rate/flux, the impacts of BVOCs on both SOA and O_3 formation should be assessed more accurately and economically in the future.

REFERENCES

Andreaniaksoyoglu, S; Keller, J. Estimates of Monoterpene and Isoprene Emissions from the Forests in Switzerland, *Journal of Atmospheric Chemistry*, 20, 1, 71-87, 1995.

Apel, EC; Riemer, DD; Hills, A; Baugh, W; Orlando, J; Faloona, I; Tan, D; Brune, W; Lamb, B; Westberg, H; Carroll, MA; Thornberry, T; Geron, CD. Measurement and interpretation of isoprene fluxes and isoprene, methacrolein, and methyl vinyl ketone mixing ratios at the PROPHET site during the 1998 Intensive, *Journal of Geophysical Research-Atmospheres*, 107, D3, 2002.

Arneth, A; Schurgers, G; Hickler, T; Miller, PA. Effects of species composition, land surface cover, CO_2 concentration and climate on isoprene emissions from European forests, *Plant Biology*, 10, 1, 150-162, 2008.

Atkinson, R. Gas-phase tropospheric chemistry of volatile organic compounds .1. Alkanes and alkenes, *Journal of Physical and Chemical Reference Data*, 26, 2, 215-290, 1997.

Atkinson, R; Arey, J. Atmospheric degradation of volatile organic compounds, *Chemical Reviews*, 103, 12, 4605-4638, 2003.

Biesenthal, TA; Wu, Q; Shepson, PB; Wiebe, HA; Anlauf, KG; Mackay, GI. A study of relationships between isoprene, its oxidation products, and ozone, in the Lower Fraser Valley, BC, *Atmospheric Environment*, 31, 14, 2049-2058, 1997.

Brown, SS; Degouw, JA; Warneke, C; Ryerson, TB; Dube, WP; Atlas, E; Weber, RJ; Peltier, RE; Neuman, JA; Roberts, JM; Swanson, A; Flocke, F; McKeen, SA; Brioude, J; Sommariva, R; Trainer, M; Fehsenfeld, FC; Ravishankara, AR. Nocturnal isoprene oxidation over the Northeast United States in summer and its impact on reactive nitrogen partitioning and secondary organic aerosol, *Atmospheric Chemistry and Physics*, 9, 9, 3027-3042, 2009.

Carter, WPL; Atkinson, R. Development and evaluation of a detailed mechanism for the atmospheric reactions of isoprene and NOx, *International Journal of Chemical Kinetics*, 28, 7, 497-530, 1996.

Chameides, WL; W Lindsay R, ; Richardson, J; Kiang, CS. The Role of Biogenic Hydrocarbons in Urban Photochemical Smog - Atlanta as a Case-Study, *Science*, 241, 4872, 1473-1475, 1988.

Chameides, WL; Fehsenfeld, F; Rodgers, MO; Cardelino, C; Martinez, J; Parrish, D; Lonneman, W; Lawson, DR; Rasmussen, RA; Zimmerman, P; Greenberg, J; Middleton, P; Wang, T. Ozone Precursor Relationships in the Ambient Atmosphere, *Journal of Geophysical Research-Atmospheres*, 97, D5, 6037-6055, 1992.

Cheng, HR; Guo, H; Saunders, SM; Lam, SHM; Jiang, F; Wang, XM; Simpson, IJ; Blake, DR; Louie, PKK; Wang, TJ. Assessing photochemical ozone formation in the Pearl River Delta with a photochemical trajectory model, *Atmospheric Environment*, 44, 34, 4199-4208, 2010a.

Cheng, HR; Guo, H; Saunders, SM; Wang, XM. Roles of volatile organic compounds in photochemical ozone formation in the atmosphere of the Pearl River Delta, southern China, *Air Quality and Climate Change*, 44, 4, 29-38, 2010b.

Cheng, HR; Guo, H; Wang, XM; Saunders, SM; Lam, SHM; Jiang, F; Wang, TJ; Ding, AJ; Lee, SC; Ho, KF. On the relationship between ozone and its precursors in the Pearl River Delta: application of an observation-based model (OBM), *Environmental Science and Pollution Research*, 17, 3, 547-560, 2010c.

Cheung, K; Guo, H; Ou, JM; Simpson, IJ; Barletta, B; Meinardi, S; Blake, DR. Diurnal profiles of isoprene, methacrolein and methyl vinyl ketone at an urban site in Hong Kong, *Atmospheric Environment*, 84, 323-331, 2014.

Ciccioli, P; Brancaleoni, E; Cecinato, A; Sparapani, R; Frattoni, M. Identification and Determination of Biogenic and Anthropogenic Volatile Organic-Compounds in Forest Areas of Northern and Southern Europe and a Remote Site of the Himalaya Region by High-Resolution Gas-Chromatography Mass-Spectrometry, *Journal of Chromatography*, 643, 1-2, 55-69, 1993.

Colman, JJ; Swanson, AL; Meinardi, S; Sive, BC; Blake, DR; Rowland, FS. Description of the analysis of a wide range of volatile organic compounds in whole air samples collected during PEM-Tropics A and B, *Analytical Chemistry*, 73, 15, 3723-3731, 2001.

Devalia, JL; Bayram, H; Rusznak, C; Calderon, M; Sapsford, RJ; Abdelaziz, MA; Wang, J; Davies, RJ. Mechanisms of pollution-induced airway disease: In vitro studies in the upper and lower airways, *Allergy*, 52, 45-51, 1997.

Ding, X; Wang, XM; Gao, B; Fu, XX; He, QF; Zhao, XY; Yu, JZ; Zheng, M. Tracer-based estimation of secondary organic carbon in the Pearl River Delta, south China, *Journal of Geophysical Research-Atmospheres*, 117, 2012.

Duan, JC; Tan, JH; Cheng, DX; Bi, XH; Deng, WJ; Sheng, GY; Fu, JM; Wong, MH. Sources and characteristics of carbonaceous aerosol in two largest cities in Pearl River Delta Region, China, *Atmospheric Environment*, 41, 14, 2895-2903, 2007.

Gryparis, A; Forsberg, B; Katsouyanni, K; Analitis, A; Touloumi, G; Schwartz, J; Samoli, E; Medina, S; Rerson, H; Niciu, EM; Wichmann, HE; Kriz, B; Kosnik, M; Skorkovsky, J; Vonk, JM; Dortbudak, Z. Acute effects of ozone on mortality from the "Air pollution and health: A European approach" project, *American Journal of Respiratory and Critical Care Medicine*, 170, 10, 1080-1087, 2004.

Guenther, A; Hewitt, CN; Erickson, D; Fall, R; Geron, C; Graedel, T; Harley, P; Klinger, L; Lerdau, M; Mckay, WA; Pierce, T; Scholes, B; Steinbrecher, R; Tallamraju, R; Taylor, J; Zimmerman, P. A Global-Model of Natural Volatile Organic-Compound Emissions, *Journal of Geophysical Research-Atmospheres*, 100, D5, 8873-8892, 1995.

Guenther, A; Geron, C; Pierce, T; Lamb, B; Harley, P; Fall, R. Natural emissions of non-methane volatile organic compounds; carbon monoxide, and oxides of nitrogen from North America, *Atmospheric Environment*, 34, 12-14, 2205-2230, 2000.

Guenther, A; Karl, T; Harley, P; Wiedinmyer, C; Palmer, PI; Geron, C. Estimates of global terrestrial isoprene emissions using MEGAN (Model of Emissions of Gases and Aerosols from Nature), *Atmospheric Chemistry and Physics*, 6, 3181-3210, 2006.

Guo, H; So, KL; Simpson, IJ; Barletta, B; Meinardi, S; Blake, DR. C-1-C-8 volatile organic compounds in the atmosphere of Hong Kong: Overview of atmospheric processing and source apportionment, *Atmospheric Environment*, 41, 7, 1456-1472, 2007.

Guo, H; Ling, ZH; Simpson, IJ; Blake, DR; Wang, DW. Observations of isoprene, methacrolein (MAC) and methyl vinyl ketone (MVK) at a mountain site in Hong Kong, *Journal of Geophysical Research-Atmospheres*, 117, 2012.

Haney, JT; Conner, TH; Li, L. Detection of ozone-induced DNA single strand breaks in murine bronchoalveolar lavage cells acutely exposed in vivo, *Inhalation Toxicology*, 11, 4, 331-341, 1999.

Hanna, SR; Russell, AG; Wilkinson, JG; Vukovich, J; Hansen, DA. Monte Carlo estimation of uncertainties in BEIS3 emission outputs and their effects on uncertainties in chemical transport model predictions, *Journal of Geophysical Research-Atmospheres*, 110, D1, 2005.

Helmig, D; Ortega, J; Guenther, A; Herrick, JD; Geron, C. Sesquiterpene emissions from loblolly pine and their potential contribution to biogenic aerosol formation in the Southeastern US, *Atmospheric Environment*, 40, 22, 4150-4157, 2006.

Helmig, D; Ortega, J; Duhl, T; Tanner, D; Guenther, A; Harley, P; Wiedinmyer, C; Milford, J; Sakulyanontvittaya, T. Sesquiterpene emissions from pine trees - Identifications, emission rates and flux estimates for the contiguous United States, *Environmental Science & Technology*, 41, 5, 1545-1553, 2007.

Hong Kong Herbarium, Check List of Hong Kong Plants 2004: Agriculture, Fisheries and Conservation Department, Hong Kong, 2004.

Hsieh, C-C; Horng, S-H; Liao, P-N. Stability of trace-level volatile organic compounds stored in canisters and tedlar bags, *Aerosol and Air Quality Research*, 3, 1, 17-28, 2003.

Hu, D; Bian, Q; Li, TWY; Lau, AKH; Yu, JZ. Contributions of isoprene, monoterpenes, beta-caryophyllene, and toluene to secondary organic aerosols in Hong Kong during the summer of 2006, *Journal of Geophysical Research-Atmospheres*, 113, 2008.

Hu, D; Bian, QJ; Lau, AKH; Yu, JZ. Source apportioning of primary and secondary organic carbon in summer PM2.5 in Hong Kong using positive matrix factorization of secondary and primary organic tracer data, *Journal of Geophysical Research-Atmospheres*, 115, 2010.

Huang, H; Ho, KF; Lee, SC; Tsang, PK; Ho, SSH; Zou, CW; Zou, SC; Cao, JJ; Xu, HM. Characteristics of carbonaceous aerosol in PM2.5, Pearl Delta River Region, China, *Atmospheric Research*, 104, 227-236, 2012.

Isebrands, JG; Guenther, AB; Harley, P; Helmig, D; Klinger, L; Vierling, L; Zimmerman, P; Geron, C. Volatile organic compound emission rates from mixed deciduous and coniferous forests in Northern Wisconsin, USA, *Atmospheric Environment*, 33, 16, 2527-2536, 1999.

Isidorov, VA; Zenkevich, IG; Ioffe, BV. Volatile Organic-Compounds in the Atmosphere of Forests, *Atmospheric Environment*, 19, 1, 1-8, 1985.

Janson, R; De Serves, C; Romero, R. Emission of isoprene and carbonyl compounds from a boreal forest and wetland in Sweden, *Agricultural and Forest Meteorology*, 98-9, 671-681, 1999.

Jiang, F; Liu, Q; Huang, XX; Wang, TJ; Zhuang, BL; Xie, M. Regional modeling of secondary organic aerosol over China using WRF/Chem, *Journal of Aerosol Science*, 43, 1, 57-73, 2012.

Lam, SHM; Saunders, SM; Guo, H; Ling, ZH; Jiang, F; Wang, XM; Wang, TJ. Modelling VOC source impacts on high ozone episode days observed at a mountain summit in Hong Kong under the influence of mountain-valley breezes, *Atmospheric Environment*, 81, 166-176, 2013.

Lau, AKH; Yuan, ZB; Yu, JZ; Louie, PKK. Source apportionment of ambient volatile organic compounds in Hong Kong, *Science of the Total Environment*, 408, 19, 4138-4149, 2010.

Leung, DYC; Wong, P; Cheung, BKH; Guenther, A. Improved land cover and emission factors for modeling biogenic volatile organic compounds emissions from Hong Kong, *Atmospheric Environment*, 44, 11, 1456-1468, 2010.

Ling, ZH; Guo, H. Contribution of VOC sources to photochemical ozone formation and its control policy implication in Hong Kong, *Environmental Science and Policy*, 38, 180-191, 10.1016/j.envsci.2013.12.004, 2014.

Lippmann, M. Health-Effects of Tropospheric Ozone - Review of Recent Research Findings and Their Implications to Ambient Air-Quality Standards, *Journal of Exposure Analysis and Environmental Epidemiology*, 3, 1, 103-129, 1993.

Miyoshi, A; Hatakeyama, S; Washida, N. Om Radical-Initiated Photooxidation of Isoprene - an Estimate of Global Co Production, *Journal of Geophysical Research-Atmospheres*, 99, D9, 18779-18787, 1994.

Nichol, J; Wong, MS. Estimation of ambient BVOC emissions using remote sensing techniques, *Atmospheric Environment*, 45, 17, 2937-2943, 2011.

Otter, LB; Guenther, A; Greenberg, J. Seasonal and spatial variations in biogenic hydrocarbon emissions from southern African savannas and woodlands, *Atmospheric Environment*, 36, 26, 4265-4275, 2002.

Placet, M; Mann, CO; Gilbert, RO; Niefer, MJ. Emissions of ozone precursors from stationary sources: a critical review, *Atmospheric Environment*, 34, 12-14, 2183-2204, 2000.

Pokorska, O; Dewulf, J; Amelynck, C; Schoon, N; Simpraga, M;. Steppe K; Van, H. Langenhove, Isoprene and terpenoid emissions from Abies alba: Identification and emission rates under ambient conditions, *Atmospheric Environment*, 59, 501-508, 2012.

Russell, A; Dennis, R. NARSTO critical review of photochemical models and modeling, *Atmospheric Environment*, 34, 12-14, 2283-2324, 2000.

Saxton, JE; Lewis, AC; Kettlewell, JH; Ozel, MZ; Gogus, F; Boni, Y; Korogone, SOU; Serca, D. Isoprene and monoterpene measurements in a secondary forest in northern Benin, *Atmospheric Chemistry and Physics*, 7, 15, 4095-4106, 2007.

So, KL; Wang, T. C-3-C-12 non-methane hydrocarbons in subtropical Hong Kong, spatial-temporal variations, source-receptor relationships and photochemical reactivity, *Science of the Total Environment*, 328, 1-3, 161-174, 2004.

Streets, DG; Bond, TC; Carmichael, GR; Fernandes, SD; Fu, Q; He, D; Klimont, Z; Nelson, SM; Tsai, NY; Wang, MQ; Woo, JH; Yarber, KF. An inventory of gaseous and primary aerosol emissions in Asia in the year 2000, *Journal of Geophysical Research-Atmospheres*, 108, D21, 2003.

Tang, JH; Chan, LY; Chan, CY; Li, YS; Chang, CC; Liu, SC; Wu, D; Li, YD. Characteristics and diurnal variations of NMHCs at urban, suburban, and rural sites in the Pearl River Delta and a remote site in South China, *Atmospheric Environment*, 41, 38, 8620-8632, 2007.

Tsui, JKY; Guenther, A; Yip, WK; Chen, F. A biogenic volatile organic compound emission inventory for Hong Kong, *Atmospheric Environment*, 43, 40, 6442-6448, 2009.

Wang, JL; Wang, CH; Lai, CH; Chang, CC; Liu, Y; Zhang, YH; Liu, S; Shao, M. Characterization of ozone precursors in the Pearl River Delta by time series observation of non-methane hydrocarbons, *Atmospheric Environment*, 42, 25, 6233-6246, 2008.

Wang, SY; Wu, DW; Wang, XM; Fung, JCH; Yu, JZ. Relative contributions of secondary organic aerosol formation from toluene, xylenes, isoprene, and monoterpenes in Hong Kong and Guangzhou in the Pearl River Delta, China: an emission-based box modeling study, *Journal of Geophysical Research-Atmospheres*, 118, 2, 507-519, 2013.

Warneck, P. Chemistry of the Natural Atmosphere, San Diego: Academic Press, 2000.

Wei, XL; Li, YS; Lam, KS; Wang, AY; Wang, TJ. Impact of biogenic VOC emissions on a tropical cyclone-related ozone episode in the Pearl River Delta region, China, *Atmospheric Environment*, 41, 36, 7851-7864, 2007.

Wong, MS; Sarker, MLR; Nichol, J; Lee, SC; Chen, HW; Wan, YL; Chan, PW. Modeling BVOC isoprene emissions based on a GIS and remote sensing database, *International Journal of Applied Earth Observation and Geoinformation*, 21, 66-77, 2013.

Yang, CX; Yang, HB; Guo, S; Wang, ZS; Xu, XH; Duan, XL; Kan, HD. Alternative ozone metrics and daily mortality in Suzhou, The China Air Pollution and Health Effects Study (CAPES), *Science of the Total Environment*, 426, 83-89, 2012.

Yuan, B; Chen, WT; Shao, M; Wang, M; Lu, SH; Wang, B; Liu, Y; Chang, CC; Wang, BG. Measurements of ambient hydrocarbons and carbonyls in the Pearl River Delta (PRD), China, *Atmospheric Research*, 116, 93-104, 2012.

Zhang, J; Wang, T; Chameides, WL; Cardelino, C; Kwok, J; Blake, DR; Ding, A; So, KL. Ozone production and hydrocarbon reactivity in Hong Kong, Southern China, *Atmospheric Chemistry and Physics*, 7, 557-573, 2007.

Zhao, J; Zhang, RY; Fortner, EC; North, SW. Quantification of hydroxycarbonyls from OH-isoprene reactions, *Journal of the American Chemical Society*, 126, 9, 2686-2687, 2004.

Zheng, JY; Shao, M; Che, WW; Zhang, LJ; Zhong, LJ; Zhang, YH; Streets, D. Speciated VOC Emission Inventory and Spatial Patterns of Ozone Formation Potential in the Pearl River Delta, China, *Environmental Science & Technology*, 43, 22, 8580-8586, 2009a.

Zheng, JY; Zhang, LJ; Che, WW; Zheng, ZY; Yin, SS. A highly resolved temporal and spatial air pollutant emission inventory for the Pearl River Delta region, China and its uncertainty assessment, *Atmospheric Environment*, 43, 32, 5112-5122, 2009b.

Zheng, JY; Zheng, ZY; Yu, YF; Zhong, LJ. Temporal, spatial characteristics and uncertainty of biogenic VOC emissions in the Pearl River Delta region, China, *Atmospheric Environment*, 44, 16, 1960-1969, 2010.

In: Volatile Organic Compounds
Editor: Khaled Chetehouna

ISBN: 978-1-63117-862-7
© 2014 Nova Science Publishers, Inc.

Chapter 5

NATURAL ORGANIC COMPOUNDS FROM THE URBAN FOREST OF THE METROPOLITAN REGION, CHILE: IMPACT ON AIR QUALITY

M. Préndez[*], K. Corada and J. Morales

Laboratorio de Química de la Atmósfera. Departamento de Química
Orgánica y Fisicoquímica. Facultad de Ciencias Químicas y
Farmacéuticas, Universidad de Chile.

ABSTRACT

Tree species emit oxygen and biogenic volatile organic compounds (BVOCs) which react in the atmosphere generating other chemical species including ozone (O_3). At the same time, trees capture particulate matter and gases (carbon dioxide, nitrogen oxides, and ozone).

Ozone is a secondary pollutant especially abundant in urban atmospheres. Santiago, Chile, is affected by high concentrations of O_3, especially in the northeast of the city and during the austral summer. Due to the aesthetic, climatic, and ecological benefits derived from trees, the Chilean government has been using them in a natural decontamination strategy especially for removing particulate matter and some gases not to mention the potential production of ozone through the emission of BVOCs.

---

[*] Corresponding author: Email: mprendez@ciq.uchile.cl.

Since BVOCs emissions are species-specific, their contribution to the photochemical reactivity in urban environment is very much related to plant biodiversity in the urban forest.

In this chapter we report a reanalysis and summary of the different works made by our group, focus on determining emission factor (EF) of isoprene and monoterpenes from around 33% of the Santiago urban forest. This study also includes the ability of urban forest species to generate O_3 through an index called the Photochemical Ozone Creation Index (POCI).

This report looks at 15 trees, exotic and native, studied at different stages of growth (small, young and adult) and seasons (austral autumn and spring). Standard EF are reported according to normalization proposed by Guenther et al., 1995. A discussion considering the standard and non standard EF is also included. Results show that, in general, exotic trees are more pollutant than native trees because the EF and POCI values of exotic trees are higher than those of native species. Those results are relevant if trees are used for decontamination purposes.

Modifications made to the emission inventory of BVOCs replacing EF, included by default in the model by the experimental EF, demonstrate that the taxonomical approaches of EF overestimate biogenic emissions. The integration of the different parameters contributes to the discussion for selecting the species more beneficial for the Metropolitan Region of Chile from the environmental and human health point of view.

Also some information is given about BVOC emissions from other Latin American countries.

Keywords: Native and non-native species, BVOCs, growth stages, Emission Factors, Photochemical Ozone Creation Index, urban forest, Metropolitan Region of Chile

1. INTRODUCTION

Very large amounts of various volatile organic compounds (VOCs) from natural and anthropogenic sources are emitted to the atmosphere. Natural sources correspond to 69% of the total emission of non-methane VOCs (NMVOCs), and 31% is attributed to the anthropogenic emission (Kansal, 2009). The biogenic NMVOCs include natural sources such as geological hydrocarbon deposits and volcanoes (Fehsenfeld et al., 1992), oceans, soils and sediments, and microbial decomposition of organic matter, but the most important is the emission from vegetation, especially forests (Guenther et al., 1995). The anthropogenic sources include, among others, petrochemical plants, vehicles, paints for industrial and commercial use, thinners, and solvents (Leung et al., 2010).

In 1960, Went (1960) first proposed that the natural emission of VOCs from the leaves of trees and vegetation could have significant effects on atmospheric chemistry; at present over 30,000 compounds have been characterized, most of them originating in terrestrial vegetation. The biogenic volatile organic compounds (BVOCs) are generally unsaturated linear and cyclic compounds dominated by isoprene (2-metil-1,3-butadieno, molecular formula C_5H_8), hemi-terpene (C_5), and monoterpenes (C_{10}, two molecules of isoprene) (IUPAC web page, 2012) which represent a significant fraction of total atmospheric BVOCs. More than 15 monoterpenes have been found in vegetable species (Sharkey and Yeh, 2001; Arneth et al., 2008); all of them represent a source of atmospheric hydrocarbons corresponding to about double the emissions from anthropogenic sources (Calfapietra et al., 2009). Many researchers have been working to determine their emission velocities (Zimmerman, 1979; Arnts and Meeks, 1981; Tingey et al., 1991; Guenther et al., 1995; and Lamb et al., 1985, 1986, 1987, 1993). Also, there are studies about their distribution and the oxidation products formed in the atmosphere (Yokouchi and Ambe, 1984; Isidorov et al., 1985; Fehsenfeld et al., 1992; Montzka et al., 1993, 1995).

BVOCs are emitted by plants to attract pollinators, act as agents of defense against pathogens and herbivore predators, communicate with other plants and organisms, or as plant membranes to protect against high temperatures (Peñuelas and Llusia, 2002). BVOCs also perform functions relating to pigments, hormones, and membrane constituents, among others. They originate in different plant tissues through various physiological processes and accumulate in leaves and stems and are emitted or stored depending on the species (Pichersky and Gershenzon, 2002). Speed of BVOCs emission is determined by the rate of synthesis, physiological and physicochemical characteristics, mainly its solubility (in water), volatility (very strong odors), and diffusivity (Peñuelas and Staudt, 2010). They constitute an important fraction of essential oils, resins, and waxes, and so provide a range of commercially useful compounds used as solvents, spices, adhesives, and flower fragrances (McGarvey and Croteau, 1995); for this reason they are also quite relevant for urban parks and gardens.

Terpenes play a crucial importance for leaf surface, ecosystems, atmospheric chemistry (Holopainen, 2004), and in the emission of greenhouse gases (CO_2, CH_4 y N_2O) (Denman et al., 2007; Monson et al., 2007). Two to 5% of the carbon assimilated by plants as CO_2 in photosynthesis is released as

BVOCs (Loreto and Schnitzler, 2010). Atmospheric concentrations of BVOCs vary over a wide range from a few ng L^{-1} to several μg L^{-1}.

Table 1 shows the main types of BVOCs issued by plant species, their reactivity, and their estimated atmospheric concentration.

The BVOCs commonly emitted from trees include isoprene and some monoterpenes such as α-pinene, β-pinene, myrcene and limonene (Nowak and Sisinni, 1993). Figure 1 shows the molecular structures of certain terpenes.

According to Tingey et al. (1991), the emissions of BVOCs are specie-specific and depend on the source within the leaf tissue, the diffusion path, the volatility of the compound, and environmental factors.

The biosynthesis of most BVOCs can be categorized into one of three major biosynthetic pathways in plants: terpene biosynthesis pathway, oxylipins biosynthesis pathway, and shikimate-benzoic acid biosynthesis pathway (Dudareva et al., 2004). Studies on the biosynthesis of terpenes present major achievements including the complete elucidation of the metileritritol pathway (MEP) responsible for one of the basic units, C5 isopentenyl diphosphate (IDP) and dimetilalildiphosphate (DMADP), in many bacteria and in the plastids of all phototropic divisions of organisms (Loreto and Schnitzler, 2010).

Table 1. Main types of BVOCs emitted by plants

Chemical compound	Estimate Global Emission, annual (Tg C)	Reactivity (half-life in the atmosphere) (h)	Atmospheric concentration	Example	Source
Isoprene	175 – 503	4.8	ng L^{-1} certain μg L^{-1}	isoprene	Almost all tree
Monoterpenes	127 – 480	2.4 – 4.8	ng L^{-1} certain μg L^{-1}	α-pinene, β-pinene	Variety pine
				limonene	Variety of citrus
Sesquiterpenes		0.03 – 0.06	Not detectable due to high reactivity	β-caryophyllene	Some trees

Source: Kesselmeier and Staudt (1999).

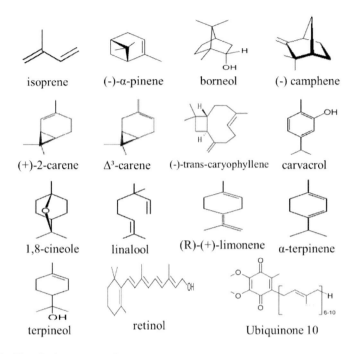

Figure 1. Chemical structure of some terpenes.

IDP and DMADP are formed through two biosynthetic pathways in plants: 1) via cytosolic mevalonic acid (MVA) from acetyl-CoA pathway, 2) Plastidic pathway MEP from pyruvate and glyceraldehyde 3-phosphate (G3P). Generally, the MEP pathway provides IDP and DMADP for the biosynthesis of isoprene and monoterpenes while the MVA pathway provides C5 units for the formation of sesquiterpenes. At present it is known that sometimes the MEP pathway may also contribute to the formation of sesquiterpenes.

According to many researchers, the most influencing environmental variables affecting BVOC emissions are temperature, relative humidity, and photosynthetically active radiation (PAR).

2. Enviromental Variables Affecting BVOC Emissions

Photosynthetically active radiation (400-700nm) and temperature are two factors that can affect the ability of BVOC emission from the leaves of plants.

Laboratory studies in plants emitting isoprene report that growth under favorable conditions of temperature and PAR increase the emission capability (Monson et al., 1992; Harley et al., 2004). For monoterpenes, the relationship between emission and these two variables is still under discussion. At least in part, this is the reason for the normalization of emission factors (EF) suggested by Guenther, et al., (1995). In Santiago, almost all species from the urban forest studied by Préndez et al. (2013 and 2013a), Morales (2013) and Corada et al. (2014) do not show this dependence.

Juuti et al. (1990) and Janson (1993) report that relative humidity does not appear to influence the emission of monoterpenes from *Pinus radiata* and *Pinus sylvestris*; the same is reported by Schade et al. (1999) and Morales (2013) for other species. However, other studies indicate that relative humidity would have a positive effect, but slight, on the emission of monoterpenes from *Eucaliptus globulus* and *Quercus ilex* (Guenther et al., 1991, and Loreto et al., 1996, respectively).

Stress (hydric or physical) affects the physiological state of plants. Hydric stress has been studied in various greenhouse plants. Under moderate stress conditions, when the exchange of CO_2 and water vapor decrease, emissions do not change; however, under conditions of high water stress, emissions decrease (Staudt et al., 1997). Physical stress is caused by manipulation of the leaves, attack of herbivores, and pathogens and may affect emissions of BVOCs in the short or in the long term. Loreto et al. (2000) reported sectioned needles of *Pinus pinea*, in extreme brightness or darkness, emitting large amounts of monoterpenes, mainly α-pinene and limonene. The author suggests that the emission of monoterpenes is a defense response of plants especially when young. However, isoprene emission not seems to be affected or inhibited by damage caused to the plant (Loreto and Sharkey, 1993).

3. CHEMICAL INTERACTIONS OF BVOCs IN THE ATMOSPHERE

The hydroxyl radical (•OH) is an essential chemical species in the atmosphere. Basically, its formation results from the photodissociation of ozone in the presence of water vapor, with the initial formation of highly energetic atomic oxygen or singulet oxygen (O^1D) that disables to triplet oxygen (O^3P), which reacts with oxygen molecules in a reaction catalyzed by particles also present. Ozone is produced in the troposphere via photolysis of nitrogen dioxide (NO_2) in a cycle that includes the formation of nitric oxide (NO) and (O^3P) which react with oxygen molecules to form ozone which, in turn, quickly oxidizes NO to NO_2 closing the cycle. In the atmosphere these photochemical reactions are in balance unless there are other species that interfere with the NO-NO_2 cycle.

Hydroxyl radicals can react in the atmosphere with hydrocarbons having a double bond, such as terpenes, primarily via their addition to it, so that the hydroxyl radical's half-life in the troposphere is relatively short compared with other organic species. Oxidation reactions of biogenic and anthropogenic hydrocarbons with •OH radicals in the atmosphere produce peroxy radicals (RO_2•) and hydroperoxy radicals (HO_2•) which rapidly oxide NO to NO_2 interfering with the consumption of O_3. Figure 2 shows these chemical reactions.

Table 2 shows the atmospheric mean lives of the reaction of some VOCs with ozone, hydroxyl and nitrate radicals. The most reactive monoterpene is the α-terpinene; other very reactive monoterpenes are linalool, limonene, α-pinene, and 3-carene. The isoprene reacts rapidly with ·OH radicals but slowly with ozone or ·NO_3 radical.

Table 2. Times of tropospheric mean life of some VOCs due to the reaction with O_3 and •OH and •NO_3 radicals

Compound	Mean life for the reaction with		
	·OH[a]	O_3[b]	·NO_3[c]
Formaldehyde	1.5 days	-	80 days
Acetaldehyde	11 h	-	17 days
Isoprene	1.7h	1.3 days	1.7 days
α-Pinene	3.4 h	4.6 h	2.0 h
(-)-Camphene	3.5 h	18 days	1.5 days
(+)-2-Carene	2.3 h	1.7 h	36 min
Δ^3-Carene	2.1 h	10 h	1.1 h
1,8-Cineole	1.4 days	>110 days	16 years
Linalool	1.2 h	52 min	3 min
(R)-(+)-Limonene	1.1 h	1.9 h	53 min
α-Terpinene	31 min	3 min	4 min

[a] for a mean concentration of ·OH (12 h day) of $1.5 * 10^6$ molecules*cm^{-3} (0.06 pg l^{-1}) (Prinn et al., 1987).

[b] for a mean concentration of O_3 (24 h day) of $7*10^{11}$ molecules* (30ngl^{-1}) (Logan, 1985).

[c] for a mean concentration of ·NO_3 (12 h day) of $2.4*10^7$ molecules*cm^{-3} (1pg l^{-1}) (Atkinson et al., 1986).

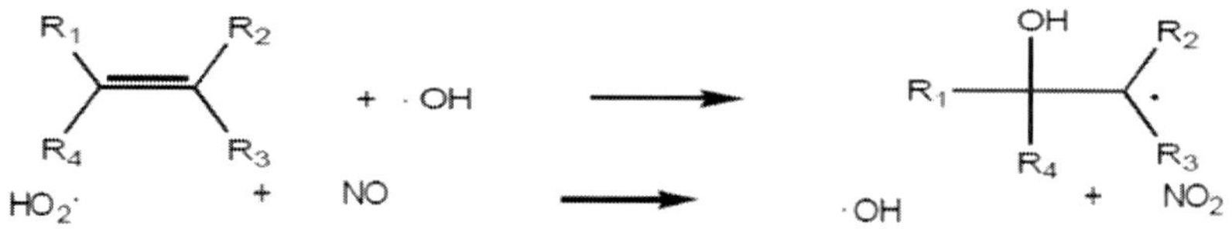

Figure 2. Chemical reactions of hydrocarbons with double chemical bond and nitrogen oxides.

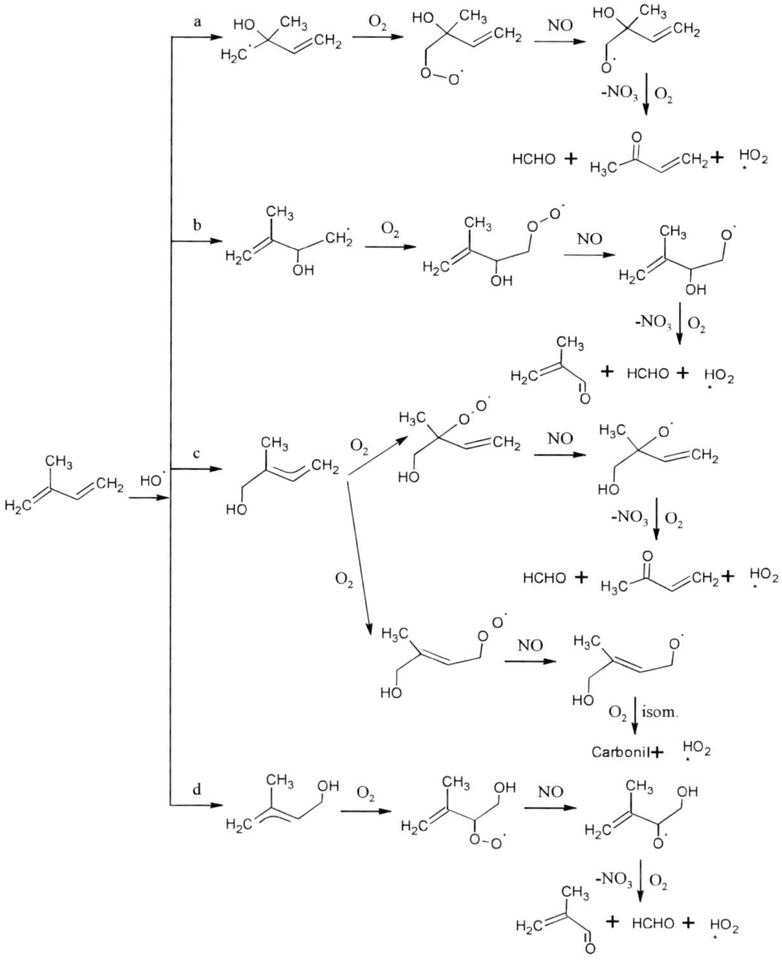

Figure 3. Mechanism of reaction between isoprene and the hydroxyl radical. Adapted from Aschmann et al. (1998), and Seinfeld and Pandis (2006).

Natural Organic Compounds from the Urban Forest ...   111

There is little information regarding the products and reaction mechanisms of terpenes with chemical species in the troposphere, even if the double bonds in the structures of terpenes makes them susceptible to the attack by radical species such as •OH, •NO$_3$, O$_3$ and other chemical species. The most studied BVOCs is the isoprene, and its major degradation products are methacrolein and methyl vinyl ketone (Finlayson-Pitts and Pitts, 2000). All terpene chemistry known to date is based on isoprene chemistry, where the Criegee biradical (Grosjean et al., 1994; Finlayson-Pitts and Pitts, 1997; Saunders et al., 2003), very unstable, is always the intermediate. The different ways of decomposition generate a series of radicals and oxygenate compounds. One possible mechanism between isoprene and the radical •OH is shown in Figure 3. The reaction of isoprene with the tropospheric O$_3$ is shown in Figure 4.

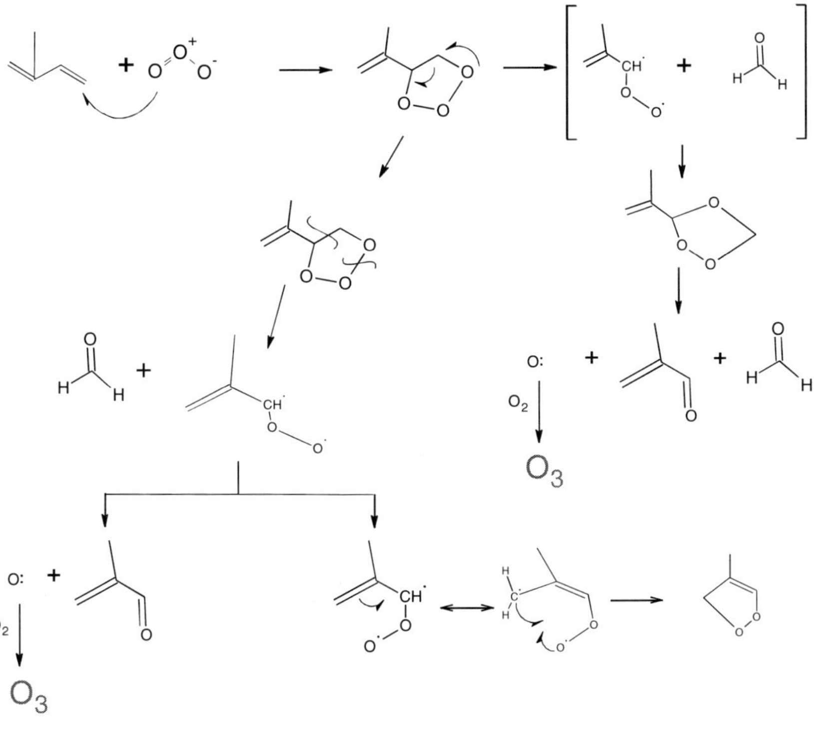

Figure 4. Mechanism of reaction between isoprene and tropospheric ozone. Adapted from Grosjean et al. (1994), Finlayson-Pitts and Pitts (1997), and Saunders et al. (2003).

The hydroperoxyl formed in a highly polluted urban atmosphere react to generate NO and the hydroxyl radical, which during the day is able to form ozone. In polluted atmospheres the nitrate radical is another potential reactant for the formation of ozone (Harley et al., 1999). Figure 5 illustrates the effect of BVOC emissions in a clean and in a polluted atmosphere.

The presence of BVOCs, such as isoprene, affects atmospheric chemistry not only generating ozone, but also secondary aerosols and other chemical species, including a toxic compound, peroxyacetyl nitrate, PAN (Figure 5); this presence also affects climate change through changes in oxidation of methane, carbon monoxide balance (Andreae and Crutzen, 1997), and the global carbon cycle balance (Kroll et al., 2006). In general, disruption of atmospheric chemistry will be greater for higher concentrations of NOx (NO + NO_2). During the day, the reactions with ·OH or ·NO_3 radicals and O_3 are of similar importance, but at night dominate the reactions with the ·NO_3 radical.

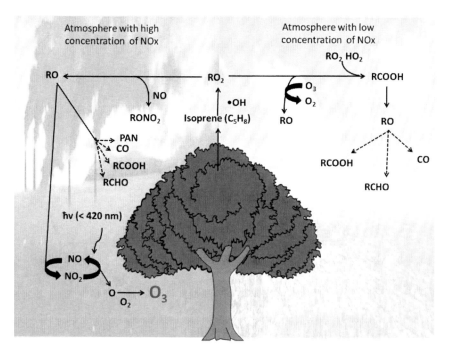

Figure 5. Reactions of isoprene in low concentrations of NOx, clean air, and high concentration of NOx, polluted air.

4. Urban Ozone and Human Health and Plants

The health problems of greatest concern from the effect due to tropospheric ozone are increased hospital admissions, exacerbation of asthma, lung inflammation, and structural alterations of the lung (WHO, 2011). The main factors affecting the magnitude of the effects are ozone concentration, time exposure, and individual sensitivity. The four most sensitive human groups are children, healthy adults who exercise outdoors, people with pre-existing respiratory diseases, and the elderly. Atmospheric pollution affects children and adults in different ways from preterm birth and intrauterine growth retardation to sudden infant death syndrome and infant mortality. People older than 65 appear to be more susceptible to respiratory problems and heart disease associated with air pollution (Cakmak et al., 2007).

In Chile, Matus and Lucero (2002) showed an increase of up to 23% in infant emergency consultations for ozone levels of around 106 $\mu g/m^3 N$ and a significant statistical relationship between daily mortality and ozone for the period 1988 – 1996 in the warmer months. Cakmak et al. (2007) showed a direct positive relationship between ozone and daily mortality and morbidity for maximum ozone concentrations around 200 $\mu g/m^3 N$ associated with a 4.9% mortality compared to 2.1% for the cold months. Epidemiological studies of time series show small positive associations between the independent effects of particulate matter and daily mortality level of 120 $\mu g/m^3 N$ ozone, 8-hour average (Chilean standard), and the value currently recommended by the World Health Organization Health Organization (WHO) which is 100 $\mu g/m^3 N$, 8-hour average. Burkart et al. (2013) find a strong pronounced effect between O_3 and mortality for a lag period of two days in Berlin and Lisbon.

Ozone enters a plant through the stomata, generating various reactive oxygen species (Mauzerall and Wang, 2001; Mills et al., 2011) and physiological effects which manifest no visible signs of damage in the short term, as the reduction of photosynthesis (Soja and Soja, 1995); an increase in the dark respiration that is associated with the need to repair and maintain the damaged tissues of the plant has also been observed. The most obvious effects are visible on leaf surfaces (chlorosis, specks, spots and necrosis), for example on tobacco, beans, watermelon, and spinach. Ozone reduces yields in crops such as carrots, tomatoes, tobacco, beans, spinach, watermelon, and wheat and other cereals. Also, some fruit trees such as peach, orange and lemon show reduction in yield (Delgado-Saborit and Esteve-Cano, 2008). But ozone also affects productivity in the long term (Faoro and Iriti, 2009; Günthardt-Georg and Vollenweider, 2007) which could be very important given that Tausz et al. (2007) also describes the effects of ozone, CO_2 and drought on plants under global change.

Sanderson et al. (2003), Pacífico et al. (2009), and Peñuelas and Staudt (2010) also studied the influence of BVOCs on climate change. A future increase of 2-3°C mean global temperature, which is predicted to occur early this century (IPCC, 2007), could increase global emissions of VOCs, and would thereby increase tropospheric ozone and methane, and modify the oxidation capacity of the atmosphere (Liakakoua et al., 2007; Lelieveld et al., 2008). Sanderson et al. (2003) found an increase in global emissions of isoprene from 484 TgC / year in the 1990s to 615 TgC/ year in the 2090s calculating an ozone increase of 10-20 ppbv in some places. The authors also noted that changes in ozone levels are closely linked to changes in surface fluxes of isoprene in regions such as the eastern U.S. or southern China. However, this effect was much smaller on the Amazon and Africa due to low levels of NOx. On the other hand, although Pacífico et al. (2009) found that atmospheric CO_2 stimulate photosynthesis, isoprene emission appears to be inhibited at elevated CO_2 concentrations and enhanced at low CO conditions. Monoterpene emissions are also likely to be influenced by CO_2, but there is less experimental evidence than for isoprene (Fowler et al., 2009).

5. Emission Factors for Biogenic Organic Compounds

As the emission of BVOCs is specie-specific, the potential of different species to form ozone depends on the EF of the different chemicals emitted. The photochemical ozone formation from NOx and BVOCs should vary according to the set of terpenes involved and their different reactivity with various chemical species present in the atmosphere (Derwent et al., 2007). Photochemical Ozone Creation Potential (POCP) calculated for each BVOC chemical species describes relative reactivity of the compounds in the atmosphere (mainly with ·OH), and therefore is not an absolute measure of ozone productivity.

In order to calculate the environmental impact of ozone formation due to the tree species studied, we relate the chemical reactivity in the atmosphere of each terpene emitted, through POCP, with the EF of terpenes emitted by each tree species studied, and we calculate an index, the Photochemical Ozone

Natural Organic Compounds from the Urban Forest ...   115

Creation Index, POCI, which accounts for both aspects and allows ranking different tree species according to their potential to generate ozone (Préndez et al., 2013).

5.1. Emission Factors for Trees of Santiago, Chile

A reanalysis of published results and some additional information on BVOCs studied in the Laboratory for Atmospheric Chemistry, Faculty of Chemical and Pharmaceutical Sciences, University of Chile, are developed in this chapter. Fifteen tree species in the urban forest, exotics and natives at different seasons, years and ages, are included.

5.1.1. Materials and Method

A sampling of nine exotic and six native species was selected. The nine exotic species were *Prunus cerasifera* (PC), *Prunus cerasifera var.nigra Pissadii* (PCVNP), *Robinia pseudoacacia* (RP), *Acacia dealbata* (AD), *Betula pendula* (BP), *Olea europaea* (OE), *Liquidambar styraciflua* (LS), *Brachychiton populneus* (BrP), and *Quercus suber* (QSu). The six native species were *Acacia caven* (AC), *Cryptocaria alba* (CA), *Schinus molle* (SM), *Quillaja saponaria* (QSa), *Caesalpinia spinosa* (CS), and *Maytenus boaria* (MB). These tree species represent around 33% of urban trees in the Metropolitan Region (MR); they grow and were in situ sampled on the North Campus of the Universidad de Chile, (33.5° Lat S y 70.6° Long W) Santiago, during austral spring and autumn and at different stages of growth.

Quercus suber, Brachychiton populneus, Liquidambar styraciflua, Quillaja saponaria and Caesalpinia spinosa were sampled as small individuals (1-2 m). *Prunus cerasifera, Prunus cerasifera var.Nigra pissardii, Quercus suber, Cryptocaria Alba, Schinus molle, Quillaja saponaria* were sampled as young individuals (>2m < 10m). *Betula Pendula, Olea Europea, Acacia delbata, Robinia pseudoacacia, Quercus suber, Cryptocaria alba, Schinus molle, Acacia caven, Maytenus boaria* and *Quillaja saponaria* were sampled as adult individuals. Additionally for *Schinus molle*, the same individual was sampled in 2005 and again in 2012.

Sampling was done using static enclosure technique. The method of quantification of isoprene and monoterpenes, data processing including the mathematical expression to calculate the EF expressed in $\mu g\ g_{ldw}^{-1}\ h^{-1}$, standardization of values for each terpene (30°C temperature and $1000\mu mol\ m^{-2}\ s^{-1}$ photosynthetically active radiation (PAR), and calculation of the corresponding POCI are presented in detail in Préndez et al. (2013). For statistical purposes, t-Student was used.

5.1.2. Results and Discussion

As was shown in previous works (Préndez et al., 2013; Préndez et al., 2013a; Morales, 2013), terpenes emitted by different trees studied on different days in different stages of growing and in different seasons, do not necessarily statistically correlate with PAR, temperature, or relative humidity. Every compound presents a different behavior; sometimes correlation was 95%, sometimes 99%, and sometimes no correlation at all was observed. Isoprene never correlates with any of the different environmental variables. These results suppose that standardization to 30°C and 1000μmol m^{-2} s^{-1} as proposed by Guenther et al. (1995), could be misleading to the reality, at least in the Metropolitan Region of Chile, where EF and POCI values in autumn result artificially increased and in spring artificially decreased. However, standardization is a good way to make an international comparison of different species and or environments. For this reason all EF reported here are standardized.

Figure 6 shows a general sheme of results for EF and POCI for the fifteen species reported at present (Préndez et al., 2013 and 2013a, Morales 2013; Corada et al., 2014). Figure 6 shows that the principal BVOC emitted from the studied species is isoprene, especially emitted from the native species. Statistical significant difference was found for total EF and monoterpenes EF between exotic and native species (p< 0.05).

A detailed study following variables such us growing and season are shown in figures 7 to 17.

5.2. Effect of the Species Growth Stage over the Emission Factors and the Photochemical Ozone Creation Index for Urban Trees in Santiago

Figure 7 shows standard EF and their corresponding standard deviation, and figure 8 shows the respective POCI obtained for the exotic and native small species. Exotic and native small trees show that EF of isoprene is always higher than EF of monoterpenes; these differences are especially important in the case of *Brachychiton populneus*. EF of monoterpenes of both small native species are similar. In all cases total EF follow the EF of isoprene, which is shown as the most important chemical species emitted. Note the big standard deviation of EF for monoterpenes from *Caesalpinea spinosa* (0.068 ± 0.14 µg g_{ldw}^{-1} h^{-1}). No statistical significant difference was found for EF between exotic and native small species. Figure 8 shows the corresponding POCI values for the small species. POCI of the native species are a little smaller than those of the exotic species.

Figure 9 shows that EF of isoprene and monoterpenes of young native species are smaller than exotic species, however no statistical significant difference was found for EF between exotic and native young species. *Cryptocaria alba* is the native tree with the smallest EF for isoprene and then EF total of all young species; EF of monoterpenes of *Cryptocaria alba* and *Quillaja saponaria* are similar. In all species EF total is essentially determined by the isoprene. This fact determines that the Photochemical Ozone Creation Index of *Cryptocaria alba* is the smallest of all exotic and native young species.

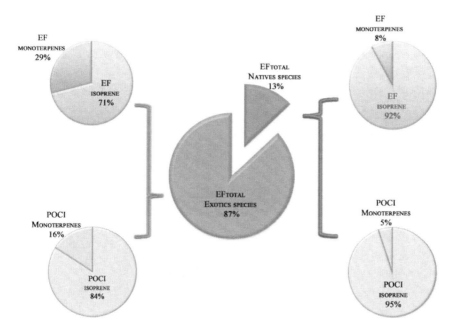

Figure 6. General results of emission factors (EF) and Photochemical Ozone Creation Potential Index (POCI) for native and exotic species.

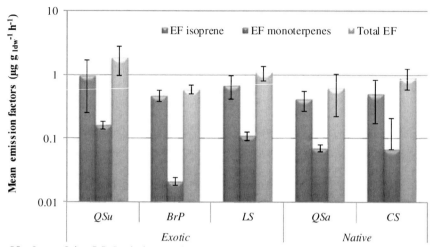

QSu: *Quercus Suber;* BrP: *Brachychiton populneus;* LS: *Liquidambar styraciflua;* QSa: *Quillaja saponaria;* CS: *Caesalpinia spinosa*

Figure 7. Standard emission factors and the corresponding standard deviation for isoprene, monoterpenes, and total for exotic and native small species of urban trees in Santiago expressed in $\mu g\ g_{ldw}^{-1}\ h^{-1}$.

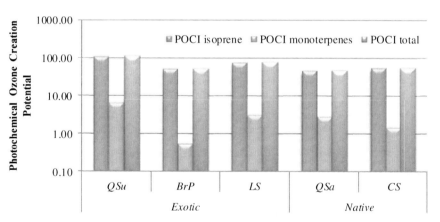

QSu: *Quercus Suber;* BrP: *Brachychiton populneus;* LS: *Liquidambar styraciflua;* QSa: *Quillaja saponaria* CS: *Caesalpinia spinosa.*

Figure 8. Photochemical Ozone Creation Index for isoprene, monoterpenes, and total for exotic and native small species of urban trees in Santiago.

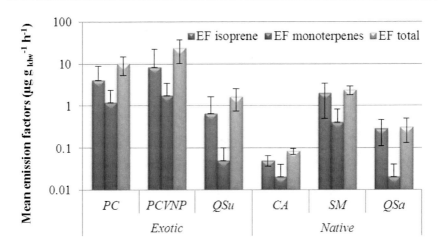

PC: Prunus cerasifera; PCVNP: Prunus cerasifera var.nigra Pissadii; QSu: Quercus suber; CA: Cryptocaria alba ; SM: Schinus molle; QSa: Quillaja saponaria

Figure 9. Standard emission factors and the corresponding standard deviation for isoprene, monoterpenes, and total for exotic and native young species of urban trees in Santiago expressed in µg g_{ldw}^{-1} h^{-1}.

Figure 10 shows the standard emission factors for adult species. It can be observe that the total EF of exotic species is always higher than the native species except for *Quercus suber*. In all species except *Betula pendula*, EF of isoprene is higher than EF of monoterpenes. Especially interesting is the small EF of monoterpenes from *Acacia caven*. At the adult stage, *Maytenus boaria* presents smaller EF than *Cryptocaria alba*. It is also interesting to note that *Quercus suber* have the smallest EF of exotic trees and is in the range of the native trees. Statistical significant difference was found for EF of monoterpenes between exotic and native adult species ($p<0.05$).

Figure 11 shows that the POCI from monoterpenes of *Acacia caven* and *Maytenus boaria* are not so different, although POCI of *Maytenus boaria* is the smallest, even if the corresponding EF are very different. This result reflects the importance of determining in detail the composition of the monoterpene mixture emitted by the tree since the impact of each monoterpene in generating ozone is different (see table 2).

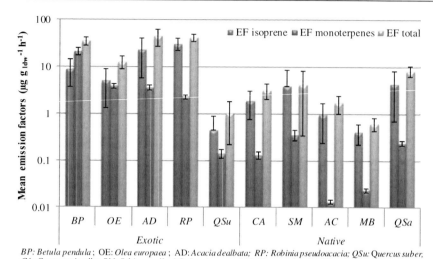

BP: *Betula pendula*; OE: *Olea europaea*; AD: *Acacia dealbata*; RP: *Robinia pseudoacacia*; QSu: *Quercus suber*; CA: *Cryptocaria alba*; SM: *Schinus molle*; AC: *Acacia caven*; MB: *Maytenus boaria*; OSa: *Quillaja saponaria*

Figure 10. Standard emission factors for isoprene, monoterpenes, and total for exotic and native adult species of urban trees in Santiago expressed in $\mu g\ g_{ldw}^{-1}\ h^{-1}$.

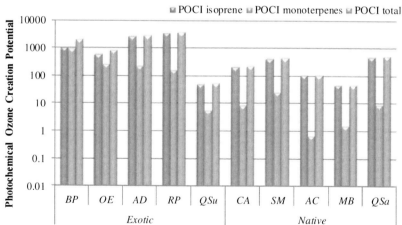

BP: *Betula pendula*; OE: *Olea europaea*; AD: *Acacia dealbata*; RP: *Robinia pseudoacacia*; QSu: *Quercus suber*; CA: *Cryptocaria alba*; SM: *Schinus molle*; AC: *Acacia caven*; MB: *Maytenus boaria*; QSa: *Quillaja saponaria*

Figure 11. Photochemical Ozone Creation Index for isoprene, monoterpenes, and total for exotic and native adult species of urban trees in Santiago.

Figure 12 presents a summary of the results obtained considering the growth stage of the species studied from the urban trees in Santiago. Even if there is some overlapping of the EF in the cases of exotic young and adult species, these last ones attain higher values. But the comparison of the POCI values clearly shows the biggest impact of the exotic adult in generating ozone. Native species are certainly, at all ages, less pollutant than the exotic species.

A special case in relation to the emissions and the growth stage of the tree is presented in figure 13. The same individual of *Schinus molle* was studied with seven years of difference. All other factors are the same regarding exposition to the sun, place and management. Figure 13 shows that EF of isoprene decreases with the age of the tree, but EF of monoterpenes increases. As isoprene emission is the most important BVOC, total EF and total POCI decrease with age (around 27%). This result does not agree with the above information. It is possible that this could be related to the decreasing of gas exchange (CO_2-O_3-BVOCs) as species pass into adulthood. So, apparently more attention should be paid to sampling emissions during the period of adulthood. Corchnoy et al. (1992) report a standard EF for *Schinus molle* (Californian pepper) of 3.7 ± 3.4 µg $g_{ldw}^{-1}h^{-1}$. This value agrees with the standard EF for the adult individual sampled in 2005.

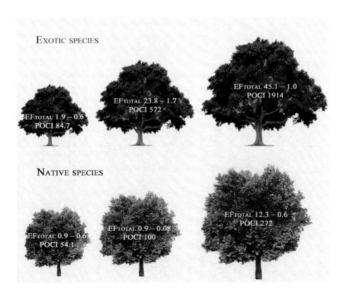

Figure 12. Range of EF and values of POCI for native and exotic species of the urban forest of Santiago, Chile expressed in µg g_{ldw}^{-1} h^{-1}.

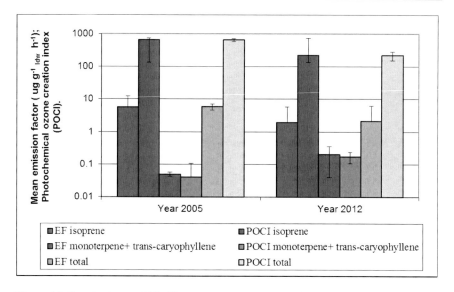

Figure 13. Standard mean EF of isoprene, monoterpenes plus trans-caryophyllene, and total EF expressed in μg g_{ldw}^{-1} h^{-1}, and the corresponding POCI and standard deviations for *Schinus molle* adult in spring with seven years age difference.

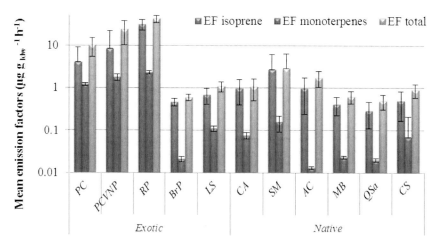

PC: *Prunus cerasifera*; PCVNP: *Prunus cerasifera var. nigra Pissadii*; RP: *Robinia pseudoacacia*; BrP: *Brachychiton populneus*; LS: *Liquidambar styraciflua*; CA: *Cryptocaria alba*; SM: *Schinus molle*; AC: *Acacia caven*; MB: *Maytenus boaria*; QSa: *Quillaja saponaria*; CS: *Caesalpinia spinosa*

Figure 14. Standard emission factors for isoprene, monoterpenes, and total for exotic and native species of urban trees in the spring of Santiago expressed in μg g_{ldw}^{-1} h^{-1}.

5.3. Effect of the Season over the Emission Factors and the Photochemical Ozone Creation Index from the Urban Trees in Santiago

Figure 14 shows standard EF for the spring season of the urban trees in Santiago. *Robinia pseudoacacia* is clearly the specie with the highest EF, determined principally by the high value of EF of isoprene. The POCI value (Figure 15) is largely higher, around 4.6 times the POCI of *Prunus* family, about 46 times the POCI of *Liquidambar styraciflua,* and about 68 times the POCI of *Brachychiton populneus*, which has the smaller EF total for exotic species and is found in the range of the native species. *Schinus molle* is the native species with the highest values of EF of isoprene and monoterpenes. The same is true for the POCI values (Figure 15). The *Maytenus boaria* is the native species with the smallest EF total. However, the smallest POCI corresponds to *Quillaja saponaria*. On the other hand, *Acacia caven* shows the smallest value for EF of monoterpenes (Figure 15). All these results relieve the importance of the knowledge of the monoterpenes mixture emitted from each tree species.

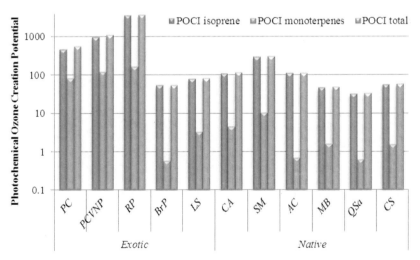

PC: *Prunus cerasifera;* PCVNP: *Prunus cerasifera var.nigra Pissadii* ; RP: *Robinia pseudoacacia;* BrP: *Brachychiton populneus;* LS: *Liquidambar styraciflua;* CA: *Cryptocaria alba;* SM: *Schinus molle;* AC: *Acacia caven;* MB: *Maytenus boaria;* QSa: *Quillaja saponaria* ; CS: *Caesalpinia spinosa*

Figure 15. Photochemical Ozone Creation Index (POCI) for isoprene, monoterpenes, and total for exotic and native species in the spring.

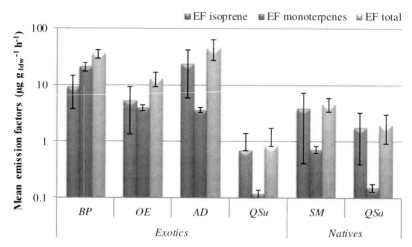

BP: *Betula pendula*; OE: *Olea europaea*; AD: *Acacia dealbata*; QSu: *Quercus suber*; SM: *Schinus molle*; QSa: *Quillaja saponaria*

Figure 16. Standard emission factors for isoprene, monoterpenes and total for exotic and native species of urban trees in the autumn of Santiago expressed in $\mu g\ g_{ldw}^{-1}\ h^{-1}$.

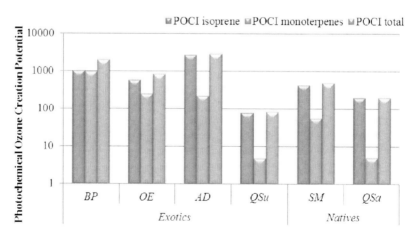

BP: *Betula pendula*; OE: *Olea europaea*; AD: *Acacia dealbata*; QSu: *Quercus suber*; SM: *Schinus molle*; QSa: *Quillaja saponaria*

Figure 17. Photochemical Ozone Creation Index for isoprene, monoterpenes and total for exotic and native species in the autumn of Santiago.

Figure 16 shows standard EF for the autumn season of the urban trees in Santiago. *Acacia dealbata* is the exotic specie with highest EF of isoprene and EF total. It is interesting to note that EF of monoterpenes of *Betula pendula* are the highest EF of monoterpenes of all studied species, exotic or native, and the highest POCI total (Figure 17) corresponds to *Acacia dealbata* essentially due to the POCI of isoprene. During autumn the less pollutant specie is *Quercus suber* followed by the native species *Schinus molle* and *Quillaja saponaria* (Figure 17).

When we compare the EF and POCI of two species *Schinus molle* and *Quillaja saponaria* sampled during the spring and autumn periods, it can be observed that *Schinus molle* shows total EF in spring is around 5.7 times higher than total EF of *Quillaja saponaria*; in the autumn the difference between these species is around 2.3 times, the major difference coming from the EF of isoprene (around 9 times higher in *Schinus molle*). t-student shows a statistical significant difference for EF of monoterpenes between exotic and native species during spring and autumn.

On the other hand, as was mentioned in materials and methods and discussed in detail in Préndez et al. (2013a), the standard EF (and POCI) are very different from those EF non- standardized, and it appeared that the differences were greater in autumn, when real EF and POCI are artificially increased with the standardization, and in spring artificially decreased in the environmental conditions of Santiago, Chile. Mean temperature are in Santiago over 30°C, during the spring-summer season, especially during central hours of the day, probably due to the heat capacity of man-made infrastrutures.

If a future global scenario of climate change is accepted, environmental changes will include an increase in temperature, higher concentrtions of CO_2, and longer drought periods (IPCC, 2007), all factors leading to increased BVOC emissions.

6. EMISSIONS INVENTORY

An emissions inventory of air pollutants is made to determine the amounts of pollutants that enter into the atmosphere from all sources in a given period of time and in a certain area. This strategy begins the process of pollution control with opportunities to effectively reduce emissions of pollutants. A full inventory, detailed and validated, accurately identifies the sources contributing the highest proportion of pollutant emissions allowing identification and instrument actions with measurable goals in terms of emission reductions. The purposes of an emissions inventory may vary according to specific needs and circumstances.

Biogenic emissions inventories are built using EF, which are an approximation of the characteristic behavior of gaseous chemical compounds of each plant species in a certain time and area (Monson et al., 1995). EF represents the emission rate of a given compound to a specific type of vegetation on the unit area or leaf biomass and is expressed as $\mu g \ m^{-2} \ h^{-1}$ with respect to leaf area or $\mu g \ ghs^{-1} h^{-1}$ relative to the mass of dry biomass (Sharkey et al., 1996). Another unit used in the inventories is $\mu g \ C \ ghs^{-1} \ h^{-1}$.

Some areas of the Metropolitan Region of Chile exceed the national standard of an 8-hour mobile average of 120 $\mu g/m^3 N$ (DS N°112, 2002), especially during summer months on a third of all days (Seguel et al., 2012) and in the northeast of Santiago (Préndez and Peralta, 2005), which concentrate the communes with the largest areas of green space: Vitacura (44%), La Reina (38.3%), Las Condes (37.3%) (Hernández, 2008).

In 2000, CONAMA-Chile, (Comisión Nacional de Medio Ambiente, at present Ministerio de Medio Ambiente, MMA) calculated an emission of 9,379 t/year of BVOCs, which corresponds to 11, 62% of total 80,682 t/year of VOCs. GloBEIS model was used. This model associates the EF by default to a list of species. EF were taxonomically assigned. If there was more than one species with the same characteristics, a value was assigned according to ecological similarities. With a species whose family does not have an assigned factor, the EF assigned correspond to a major vegetation group. Different species of the same family have the same EF. The estimation for 2010, also based on taxonomic associations, was 19,248 t/year (Puente, 2010). The modeled results correspond to 19.1% isoprene, 26.1% monoterpenes, and 54.8% for other BVOCs (Figure 18).

Experimental results obtained by Préndez et al. (2013) show that these approaches may be incorrect, at least for isoprene and monoterpenes emissions, because the EF depend on the physiological and environmental conditions of tree species. A better approximation should be obtained if experimental EF are introduced in the model. Table 3 compare the spring EF experimentally determined (Préndez et al., 2013 and 2013a; Morales, 2013; Corada et al., 2014) for isoprene and monoterpenes plus the sesquiterpene trans-caryophyllene and their respective standard deviations, and the EF used in the inventory. EF obtained by taxonomical similarities (inventory) are much greater than those determined experimentally; some examples: EF of isoprene for *Maytenus boaria* incorporated into the inventory is 7.5 x10^3 times the

Natural Organic Compounds from the Urban Forest ...   127

experimental EF; EF of monoterpenes for *Acacia caven* incorporated into the inventory is $3,9 \times 10^5$ times greater than the experimental value.

Table 3. Emission factors of BVOCs determined experimentally and the corresponding values in the inventory of VOCs, 2010, expressed as µg C ghs^{-1} h^{-1}

Scientific Name	Association in the inventory	EF isoprene	EFi/ EFe	EF monoterpenes	EFi/ EFe
Robinia pseudoacacia (E)	*Robinia*	27.6 ± 7.6 (e)	2.2 x10^2	1.84 ± 0.2 (e)	46.2
		5950 (i)		85 (i)	
Prunus cerasifera var. pissardii (E)	*Prunus*	7.5 ± 12 (e)	5.7	1.4 ± 0.2 (e)	30.4
		42.5 (i)		42.5 (i)	
Prunus cerasifera (E)	*Prunus*	3.6 ± 4.2 (e)	11.8	0.97 ± 0.1 (e)	43.8
		42.5 (i)		42.5 (i)	
Betula pendula (E)	*Betula (birch)*	8.01 ± 4.7 (e)	5.3	14.6 ± 3.1 (e)	5.8
		42.5 (i)		85 (i)	
Acacia dealbata (E)	*Acacia*	20.9 ± 15.80 (e)	3.8	2.68 ± 0.3 (e)	8.8x10^2
		79.3 (i)		2380 (i)	
Olea europaea (E)	*Paulownia*	4.55 ± 3.41 (e)	9.34	2.92 ± 0.4 (e)	14.6
		42.5 (i)		42.5 (i)	
Schinus molle (N)	*Cotinus*	3.86 ± 2.7 (e)	11	0.144 ± 0.14 (e)	3 x10^2
		42.5 (i)		42.5 (i)	
Cryptocarya alba (N)	*Persea*	1.66 ± 1.01 (e)	25.6	0.085 ± 0.025 (e)	3 x10^4
		42.5 (i)		255 (i)	
Acacia caven (N)	*Acacia*	0.86 ± 0.65 (e)	92.2	0.006 ± 0.001 (e)	3.9x10^5
		79.3 (i)		2380 (i)	
Maytenus boaria (N)	Scrub woodland	0.36 ± 0.17 (e)	7.5x10^3	0.016 ± 0.002(e)	2.2x10^4
		2700 (i)		349 (i)	
Liquidambar styraciflua (E)	*Liquidambar*	0.607 ± 0.247(e)	7.5x10^3	0.11 ± 0.015(e)	2.2x10^4
		29750(i)		1275(i)	

(E): exotic specie; (N): native specie; (e): experimental; ^{i}: emission inventory.

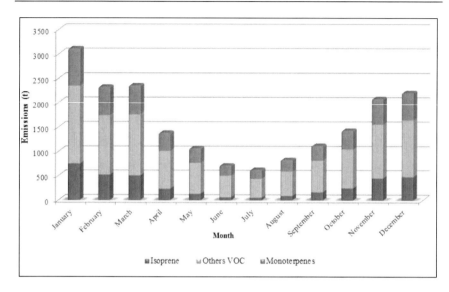

Figure 18. Emissions of VOCs estimated from GloBEIS model using EF assigned by default or by taxonomic approach (data from Puente, 2010).

Table 4. Total monthly emissions of isoprene and monoterpenes using EF determined from the inventory, 2010, and replacing some EF by those determined experimentally

Month	Total Emissions (t)	
	Inventory	Experimental
January	1509	819
Febreary	1102	595
March	1102	594
April	605	322
May	433	227
June	210	140
July	184	123
Agust	239	171
September	474	252
October	629	335
November	977	527
December	1035	559
Total	**8700**	**4659**

Another difference observed in the case of *Acacia dealbata* and *Acacia caven* is that they are assigned just to acacia, but the differences are 3.8 and 92.2 for EF from inventory and experimental EF; in the case of *Prunus cerasifera var. pissardii* and *Prunus cerasifera* they are assigned just to *Prunus* with differences of 5.7 and 11.8, respectively. From these and other differences in the assignation of EF, new calculations were done. The model does not accept many modifications simultaneously; table 4 shows final emissions resulting from five modifications to the GloBEIS introducing the determined experimental EF of isoprene and monoterpenes.

Results show that 8.700 t/year calculated by the inventory exceeds 53.6% the annual emission resulting with the introduction of the experimental EF from fifteen tree species (~ 33% of urban trees in the MR); these changes to inventory show the overestimation of BVOCs considered from native vegetation and at the same time an important reduction of the total emissions of VOCs. These results should be taken into account when strategies and actions are designed to improve air quality.

7. IMPROVING AIR QUALITY IN SANTIAGO, CHILE

The formation of ozone in cities is due, at least in part, to the presence of urban vegetation. Santiago is clearly deficient in vegetation in a large part of the city.WHO recommends $9m^2$/inhabitant of green space for urban areas. In 2003 CONAMA reported a mean value of 3.2 m^2/inhabitant for the poorest communes (range 2.9 to 0.4 m^2/inhab) and between 6.7 to 18.8 for m^2/inhab for the higher income communes (Nilo, 2003). In 2009 the mean value reported was 3.9 m^2/inhabitant with extreme values of 1.1 and 12.6 m^2/inhab (Reyes and Figueroa, 2010).

Results shown in this chapter clearly evidence that most of the native species studied emit lower concentrations of potentially ozone-forming chemical compounds, have much lower EF, and have much lower POCI. Préndez et al. (2013 and 2013a) also compared the EF of exotic species in local environmental conditions with EF in their places of origin, and again differences were found that make the exotic species more pollutant.

There is no doubt that in Santiago a good plan of planting trees is imminent; the positive effects of proper management produce correct drainage pipes preventing floods, manage soil retention, and regulate temperature, among others. It also creates a social interaction in green areas by strengthening community attachment and positive effects on human health.

At the same time, native trees, mostly evergreen, have better qualities, such as roughness, surface villi or resin, for retention of particulate matter. They have complex structures, dense foliage, and are adapted to the natural environment; all these characteristics help in capturing particulate matter, the other contaminant with high negative impact on the city of Santiago especially in winter (Préndez et al., 2011) when most exotic trees lose foliage.

In order to effectively contribute to improving the air quality in Santiago, it is urgent to make an ordered list of species according to minimal ability to impact atmospheric chemistry and, at the same time, obtain all the positive environmental effects reported for trees in the Metropolitan Region of Chile.

8. BVOCs INFORMATION FROM VARIOUS LATIN AMERICAN COUNTRIES

Some Latin American countries such as Brazil, Argentina, and Colombia report studies of BVOC emissions and their interaction with the atmosphere, specifically the reaction of ozone and the contribution to the formation of secondary organic aerosols (Yañez, 2011).

Most research related to BVOCs in Latin America are associated with the role played by these compounds in tropospheric chemistry, especially the oxidative capacity of the atmosphere and the variety of species emitted and their different reactivities (half-lives of seconds to hours). Studies agree that 80% to 90% of emissions correspond to isoprene vegetation which highly depends on temperature and sunlight. According to Harley et al. (2004), at a temperature over 30°C, 1-2% of the carbon fixed by photosynthesis is re-emitted as isoprene, which means that hot environments may make a significant contribution to isoprene. Table 6 summarizes information on some countries.

Some studies show measurements of NMVOC and other reactive VOCs (ORVOC) but agree that the main compound is isoprene, and the most mentioned monoterpenes are α-pinene, β-pinene, limonene, and 3-carene. This is also the case in Santiago. These terpenes react in the atmosphere with mean life of few hours with hydroxyl and nitrate radicals and ozone.

The studies focus mainly on working with two methods: 1) the method of static chamber studying individual species such as *Mangifera indica L.* in Brazil, *Nothofagus pumilio* and *Nothofagus antárctica* in Argentina, *Euterpe precatoria*, *Nephelium ramboutanake* in Costa Rica, *Populus fremontil* and *Pinus ponderosa* in Mexico and the fifteen species analyzed in this chapter in

Santiago, Chile; 2) the method of canopy used mostly in Amazonian studies which correspond to the Large Scale Biosphere-Atmosphere Experiment project in Amazonia (LBA), an international research conducted between 1995 and 2005 that includes countless papers with information about BVOCs, but also about weather, ecology, biogeochemistry and hydrology of Amazonia and the interactions of this area with Earth (Avissar et al., 2002).

Table 6. BVOC emissions in various Latin American countries

Country	Objective	Method	Place	Reference
Brazil	Measured sesquiterpene concentrations	Forest canopy - PTR-MS*	Central Amazonia, 60 km NNW of the city of Manaus, Brazil	Yañez (2011)
	Measured the atmospheric mixing ratios of different species of VOC	Canopy-DNPH**-coated cartridges for carbonyls and cartridges with graphitic carbons	Station at Balbina, Amazonia	Kesselmeier et al. (2000)
	Measured isoprene (*Mangifera indica L.*)	Four different enclosure systems	Brazilian Amazon (6 place different)	Harley et al. (2004)
	Measured isoprene and monoterpenes	Airborne and model (MEGAN)***	Manaus in Central Amazonia	Karl et al. (2007)
Argentina	A study was conducted to identify the basal emission of terpenes under field conditions from three *Nothofagus* species	Each species at individual sites in 0.5 dm^3 containers	San Carlos de Bariloche, Patagonia, Argentina	Centritto et al. (2008)
Colombia	First estimation of the biogenic emissions For 4 plants.	Coefficients determined from the best correlation laboratory	Aburrá Valley (Colombia)	Toro et al. (2001)

Table 6. (Continued)

Country	Objective	Method	Place	Reference
		measurements.		
Costa Rica	Two species in the family *Mimosaceae* for emitted significant quantities of isoprene and non-isoprene	Canopy Collected by relaxed eddy accumulation (REA)	La Selva Biological Station in Tropical Sarapiqui, province of Heredia	Geron et al. (2002)
	Isoprene, α-pinene, β-pinene, 3-carene, d-limonene and γ-terpinene	Cartridges filled with graphitized carbon and Tenax TA and analyzed by GC-FID****	Tropical forest in Monteverde	Esquivel-Hernández et al. (2011)
México	Emission rates from *Populus fremontil* and *Pinus ponderosa* for carbonyl compound and hydrocarbons	Dynamic enclosure system was used to collect the plant emissions using a Tedlar bags	Socorro, located in Central New México	Villanueva-Fierro et al. (2004)
Chile	Normalized experimental emissions factors (EF) for BVOCs and calculation of Photochemical Ozone Creation Index (POCI)	Static chamber, GC-FID	Santiago, Chile	Préndez et al. (2013, 2013a), Morales (2013), Corada (2014)
	Calculation of differences of ozone, NOx and	Eight monitoring	Santiago of Chile	Seguel et al. (2012)
	VOCs concentrations in Santiago	stations (MACAM-2)		

[*] Proton-Transfer Reaction Mass Spectrometer [**] Model of Emissions of Gases and Aerosols from Nature[***] Gas Cromatograph - Flame Ionization Detector.

In general, the incipient study of BVOCs in Latin America shows the necessity of better measuring and modelling BVOC emissions, in particular the isoprene emissions considering their important responsibility in the acidification of the atmosphere mentioned by Geron et al. (2002), and shown in this chapter. It is necessary to consider the role of BVOCs associated with tropospheric chemistry and its link with the ·OH radical, O_3, and the ·NO_3 radical, their concentrations at local and regional level, as well as biological factors.

POCI calculations are a first attempt to consider the potential impact of BVOCs on the atmosphere, but until now these have been reduce to considering ·OH radical interactions. More information is necessary in relation to reactions in the atmosphere between terpenes and ozone and the ·NO_3 radical. These considerations could be relevant to improving the air in cities of countries in the Latin American region which are under rapid development. At a global level, much more studies are necessary in the tropics where the radiation is more intense, the vegetation is more abundant, and so too are the emissions.

REFERENCES

Andreae M. O. & Crutzen P. J. (1997). Atmospheric aerosols: Biogeochemical sources and role in atmospheric chemistry, *Science, 276*, 1052-1058.

Arneth, A., Monson, R. K., Schurgers, G., Niinemets, U. & Palmer, P. I. (2008). Why are estimates of global terrestrial isoprene emissions so similar (and why is this not so for monoterpenes)? *Atmospheric Chemistry and Physics, 8*, 4605 – 4620.

Arnts, R. R. & Meeks, S. A. (1981). Biogenic hydrocarbon contribution to the ambient air of selected areas. *Atmospheric Environment, 15*, 1643 – 1651.

Aschmann, S., Reissell, A., Atkinson, R. & Arey, J. (1998). Products of the gas phase reactions of the OH radical with α- and β-pinene in the presence of NO. *Journal of Geophysical research, 103*, NO. D19, 25553-25561.

Atkinson, R., Winer, A. & Pitts, J. (1986). Estimation of night-time N_2O_5 concentrations from ambient NO_2 and NO_3 radical concentrations and the role of N_2O_5 in night-time chemistry. *Atmospheric Environment, 20*, 331-339.

Avissar, R., Silva Dias, P., Silva Dias, M. & Nobre, C. (2002). The Large-Scale Biosphere-Atmosphere Experiment in Amazonia (LBA): Insights and future research needs, *J. Geophys. Res., 107*, (D20), 8086.

Burkart, K., Canário, P., Breitner, S., Schneider, A., Scherber, K., Andrade, H., Alcoforado, M. J. & Endlicher, W. (2013). Interactive short-term effects of equivalent temperature and air pollution on human mortality in Berlin and Lisbon. *Environmental Pollution, 183*, 54-63.

Cakmak, S., Dales, R. E. & Blanco, V. C. (2007). Air pollution and mortality in Chile: susceptibility among the elderly. *Environmental Health Perspectives, 115*, 524–527.

Calfapietra, C., Fares, S. & Loreto, F. (2009). Volatile organic compounds from Italian vegetation and their interaction with ozone. *Environmental Pollution, 157*, 1478 – 1486,

Centritto, M., Di Bella, C., Baraldi, R., Beget, M. E., Kemerer, A., Rapparini, F., Oricchio, P., Rebella, C. & Loreto, F. (2008). Monoterpene emissions from three Nothofagus species in Patagonia, Argentina. *Journal of Plant Interactions, 3*, 119-125.

Corada, K., Morales, J. & Préndez, M. (2014). Arborización de Santiago de Chile: elección de especies para mejorar la calidad del aire. *Tecnología en Marcha, Ed. Instituto Tecnológico de Costa Rica, Special Issue*, in press. (in spanish).

Corchnoy, A., Arey, J. & Atkinson, R. (1992). Hydrocarbon emissions from twelve urban shade trees of the Los Angeles, California, air basin, *Atmospheric Environment, 26B*, 339-348.

Delgado-Saborit, J. M. & Esteve-Cano, V. J. (2008). Assessment of tropospheric ozone effects on citrus crops using passive samplers in a western Mediterranean area. *Agriculture, Ecosystems and Environment, 124*, 147–153.

Denman, K. L., Brasseur, G., Chidthaisong, P. Ciais, P. M., Cox, R. E., Dickinson, D., Hauglustaine, C., Heinze, E., Holland, D., Jacob, U., Lohamm, S., Ramachandran, P. L., Da Silva Diaz, S. C., Wofsy & Zhang, X. (2007). Couplings between Changes in the Climate System and Biogeochemistry. In: *Climate Change 2007: The Physical Science Basis. Contribution of Working Group I to the Fourth Assessment Report of the Intergovernmental Panel on Climate Change*. Solomon, S., Qin, D., Manning, M., Chen, Z., Marquis, M., Averyt, K. B., Tignor, M. & Miller, H. L. (Eds.). Cambridge University Press, Cambridge.

Derwent, R. G., Jenkin, M. E., Passant, N. R. & Pilling, M. J. (2007). Photochemical ozone creation potentials (POCPs) for different emission sources of organic compounds under European conditions estimated with a Master Chemical Mechanism. *Atmospheric Environment, 41*, 2570- 2579.

Dudareva, N., Pichersky, E. & Gershenzon, J. (2004). Biochemistry of plant volatiles. *Plant Physiology, 135*, 1893 – 1902.

Esquivel-Hernández, G., Madrigal-Carballo, S., Alfaro-Solís, R., Sibaja-Brenes, J. P. & Valdés-González, J. (2011). First Measurements of Biogenic Hydrocarbons in Air in a Tropical Cloudy Forest, Monteverde, Costa Rica, *J. Chem. Eng.*, *5*, 1097-1106.

Faoro, F. & Iriti, M. (2009). Plant cell death and cellular alterations induced by ozone: Key studies in Mediterranean conditions. *Environmental Pollution*, *157*, 1470–1477.

Fehsenfeld, F., Calvert, J., Fall, R., Goldan, P., Guenther, A. B., Hewitt, C. N., Lamb, B., Liu, S., Trainer, M., Westberg, H. & Zimmerman, P. R. (1992). Emission of volatile organic compounds from vegetation and the implications for atmospheric chemistry. *Global Biogeochem. Cycles*, *6*, 389 – 430.

Finlayson-Pitts, B. & Pitts, Jr. J. N. (1997). Tropospheric Air Pollution: Ozone, Airborne Toxics, Polycyclic Aromatic Hydrocarbons. & Particles. *Science*, *276*, 1045-1051.

Finlayson-Pitts, B. & Pitts, Jr. J. N. (2000). Chemistry of the upper and lower atmosphere. Theory, experiments and applications. *Academic Press, San Diego, California, USA*, 969 pages.

Fowler, D., Pilegaard, K., Sutton, M. A., Ambus, P., Raivonen, M., Duyzer, J., Simpson, D., Fagerli, H., Fuzzi, S., Schjoerring, J. K., Granier, C., Neftel, A., Isaksenm, I. S. A., Laj, P., Maione, M., Monks, P. S., Burkhardt, J., Daemmgen, U., Neirynck, J., Personne, E., Wichink-Kruit, R., Butterbach-Bahl, K., Flechard, J., Tuovinen, P., Coyle, M., Gerosa, G., Loubet, B., Altimir, N., Gruenhage, L., Ammannl, C., Cieslik, S., Paoletti, E., Mikkelsen, T. N., Ro-Poulsen, H., Cellier, P., Cape, J. N., Horvath, L., Loreto, F., Niinemets, U., Palmer, P. I., Rinne, J., Misztal, P., Nemitz, E., Nilsson, D., Pryor, S., Gallagher, M. W., Vesala, T., Skiba, U., Brüggemann, N., Zechmeister-Boltenstern, S., Williams, J., O'Dowdap, C., Facchini, M. C., DeLeeuw, G., Flossman, A., Chaumerliac, N. & Erisma, J. W. (2009). Atmospheric composition change: Ecosystems–Atmosphere interactions. *Atmospheric Environment*, *43*, 5193–5267.

Geron, C., Guenther, A., Greenberg, J., Loescher, H. W., Clark, D. & Baker, B. (2002). Biogenic volatile organic compound emissions from a lowland tropical wet forest in Costa Rica, *Atmospheric Environment*, *36*, 3793–3802.

Grosjean, D., Grosjean, E. & Williams, II E. L. (1994). Atmospheric chemistry of olefins: a product study of the ozone-alkene reaction with cyclohexane added to scavenge OH. *Environ. Sci. Technol.*, *28*, 186-196.

Guenther, A., Hewitt, A. C. N., Erickson, D., Fall, R., Geron, C., Graedel, T., Harley, P., Klinger, L., Lerdau, M., McKay, W. A., Pierce, T., Scholes, B., Steinbrecher, R., Tallamraju, R., Taylor, J. & Zimmerman, P. (1995). A global model of natural volatile organic compound emissions. *J. Geophys. Res.*, *100*, 8873 – 8892.

Guenther, A. B., Monson, R. K. & Fall, R. (1991). Isoprene and Monoterpene Emission Rate Variability: Observations with *Eucalyptus* and Emission Rate Algorithm Development. *J. Geophys. Res.*, *96* (D6), 10799-10808.

Günthardt-Georg, M. S. & Vollenweider, P. (2007). Linking stress with macroscopic and microscopic leaf response in trees: new diagnostic perspectives. *Environmental* Pollution, *147*, 467–488.

Harley, P., Monson, R. & Lerdau, M. (1999). Ecological and evolutionary aspects of isoprene emission from plants, *Oecologia*, *118*, 109-123.

Harley, P., Vasconcellos, P., Vierling, L., Cleomir De S. Pinheiros, C., Greenberg, J., Guenther, A., Klinger, L., Soares De Almeida, S., Neill, D., Baker, T., Phillips, O. & Malhi, Y. (2004). Variation in potential for isoprene emissions among Neotropical forest sites, *Global Change Biology*, *10*, 630–650.

Hernández, H. J. (2008). La situación del arbolado urbano en Santiago. *Revista de Urbanismo*, *18*, ISSN 0717-5051 (in Spanish).

Holopainen, J. (2004). Multiple functions of inducible plant volatiles. *Trends in Plant Science*, *9*, 529-533.

IPCC. (2007). 2007: Climate Change The Physical Science Basis. Contribution of Working Group I to the Fourth Assessment Report of the Intergovernmental Panel on Climate Change. Solomon, S., D. Qin, M. Manning, Z. Chen, M. Marquis, K. B. Averyt, M. Tignor and H. L. Miller (Eds.). *Cambridge University Press, Cambridge, United Kingdom and New York, NY, USA*.

Isidorov, V. A., Zenkevich, I. G. & Ioffe, B. V. (1985). Volatile organic compounds in the atmosphere of forest. *Atmospheric Environment*, *19*, 1 – 8.

IUPAC web page, Gold Book, Terpenes [in line]http://goldbook.iupac.org/ T06278. html [Consultation, 13. 11. 2013]

Janson, R. W. (1993). Monoterpene emissions from Scots Pine and Norwegian Spruce. *J. Geophys. Res.*, *98* (D2), 2839-2850.

Juuti, S., Arey, J. & Atkinson, R. (1990). Monoterpene Emission Rate Measurements from a Monterey Pine. *J. Geophys. Res.*, *95* (D6), 7515-7519.

Kansal, A. (2009). Sources and reactivity of NMHCs and VOCs in the atmosphere: a review. *J. Hazard. Mater.*, *16*, 17-26.

Karl, T., Guenther, A., Yokelson, R. J., Greenberg, J., Potosnak, M., Blake, D. R. & Artaxo, P. (2007). The tropical forest and fire emissions experiment: Emission, chemistry, and transport of biogenic volatile organic compounds in the lower atmosphere over Amazonia, *Journal of Geophysical Research*, *112*, (D1), 8302.

Kesselmeier, J. & Staudt, M. (1999). Biogenic Volatile Organic Compounds (VOC): An Overview on Emission, Physiology and Ecology. *Journal of Atmospheric Chemistry*, *33*, 23-88.

Kesselmeier, J., Kuhn, U., Wolf, A., Andreae, M. O., Ciccioli, P. Brancaleoni, E., Frattoni, M., Guenther, A., Greenberg, J., De Castro Vasconcellos, P., Telles de Oliva, Tavares, T. & Artaxo, P. (2000). Atmospheric volatile organic compounds (VOC) at a remote tropical forest site in central Amazonia, *Atmospheric Environment*, *34*, 4063-4072.

Kroll, J., Ng, N. Murphy, S., Flagan, R. & Seinfeld, J. (2006). Secondary organic aerosol formation from isoprene photooxidation. *Environmental Science and Technology*, *40*(6), 1869–1877.

Lamb, B., Guenther, A., Gay, D., Westberg, H. & Pierce, T. (1993). A Biogenic Hydrocarbon Emissions Inventory for the USA using a Simple Forest Canopy Model. *Atmospheric Environment*, *27*A, 1673 – 1690.

Lamb, B., Guenther, A., Gay, D. & Westberg, H. (1987). A national inventory of biogenic hydrocarbon emissions. *Atmospheric Environment*, *21*, 1695-1705.

Lamb, B., Westberg, H. & Allwine, G. (1985). Biogenic hydrocarbon emissions from deciduos and coniferous trees in the United States. *Journal of Geophysical Research*, *90*(D1), 2380 – 2390.

Lamb, B., Westberg, H. & Allwine, G. (1986). Isoprene emission fluxes determined by an atmospheric tracer technique. *Atmospheric Environment*, *20*, 1 – 8.

Lelieveld, J., Butler, T. M., Crowley, J., Dillon, T., Fischer, H., Ganzeveld, L., Harder, H., Lawrence, M. G., Martinez, M. & Taraborrelli, D. (2008). Atmospheric oxidation capacity sustained by a tropical forest. *Nature*, *452*, 737–740.

Leung, D. Y. C., Wong, P., Cheung, B. K. H. & Guenther, A. (2010). Improved land cover and emission factors for modeling biogenic volatile organic compounds emissions from Hong Kong. *Atmospheric Environment*, *44*, 1456 -1468.

Liakakoua, E., Vrekoussisa, M., Bonsangb, B., Donousisa, Ch., Kanakidoua, M. & Mihalopoulos, N. (2007). Isoprene above the Eastern Mediterranean: seasonal variation and contribution to the oxidation capacity of the atmosphere. *Atmospheric Environment*, *41*, 1002-1010.

Logan, J. (1985). Tropospheric ozone: seasonal behaviour, trends, and anthropogenic influence. *Journal of Geophysics Research*, *90*, 10463-10482.

Loreto, F. & Sharkey, T. D. (1993). On the relationship between isoprene emission and photosynthetic metabolites under different environmental conditions, *Planta*, *189*, 420–424.

Loreto, F. & Schnitzler, P. (2010). Abiotic stresses and induced BVOCs. *Trends in Plant Science*, *15*(3), 154-166.

Loreto, F., Ciccioli, P., Cecinato, A., Brancaleoni, E., Frattoni, M. & Tricoli, D. (1996). Influence of environmental factors and air composition on the emission of α-pinene from *Quercus ilex* leaves. *Plant Physiology*, *110*, 267-275.

Loreto, F., Nascetti, P., Graverini, A. & Mannozzi, M. (2000). Emission and content of monoterpenes in intact and wounded needles of the Mediterranean Pine, *Pinus pinea*. *Functional Ecology*, *14*, 589-595.

Matus, P. & Lucero, R. (2002). Norma Primaria de Calidad del Aire. *Revista Chilena de Enfermedades Respiratorias*, *18*(2), 112-122. (in Spanish).

Mauzerall, D. & Wang, X. (2001). Protecting agricultural crops from the effects of tropospheric ozone exposure: reconciling Science and Standard Setting in the United States, Europe, and Asia. *Annual Review Energy Environment*, *26*, 237-268.

Mc Garvey, D. J. & Croteau, R. (1995). Terpenoid metabolism. *The Plant Cell*, *7*, 1015-1026.

Mills, G., Pleijel, H., Braun, S., Büker, P., Bermejo, V., Calvo, E., Danielsson, H., Emberson, L., González, I., Grünhage, L., Harmens, H., Hayes, F., Karlsson, P. & Simpson, D. (2011). New stomatal flux-based critical levels for ozone effects on vegetation. *Atmospheric Environment*, *45*, 5064-5068.

Monson, R. K., Lerdau, M. T., Sharkey, T. D, Schimel, D. S. & Fall, R. (1995). Biological aspects of constructing volatile organic compound emissions inventories. *Atmospheric Environment*, *29*, 2989-3002.

Monson, R. K., Jaeger, C. H., Adams III, W. W. Diggers, E. M., Silver, G. M. & Fall, R. (1992). Relationships among isoprene emission rate, photosynthesis, and isoprene synthase activity as influenced by temperature. *Plant Physiology*, *98*, 1175 – 1180

Monson, R. K., Trahan, N., Rosenstiel, T. N., Veres, P., Moore, D., Winkinson, M., Norby, R. J. & Volder, A. (2007). Isoprene emission from terrestrial ecosystems in response to global change: minding the gap between models and observations. *Phil. Trans. R. Soc. A.*, *365*, 1677-1695.

Montzka, S. A., Trainer, M., Angevine, W. M. & Fehsenfeld, F. C. (1995). Measurements of 3-Methyl Furan, Methyl Vinyl Ketone, and Methacrolein at a Rural Forested Site in the Rural Troposphere. *J. Geophys. Res.*, *100*, 11393-11401.

Montzka, S. A., Trainer, M., Goldan, P. D., Kuster, W. C. & Fehsenfeld, F. C. (1993). Isoprene and its Oxidation Products, Methyl Vinyl Ketone and Methacroleine, in Rural Troposphere. *J. Geophys. Res.*, *98*, 1101-1111.

Morales, J. (2013). Estudio de las emisiones de terpenos por la especie nativa Schinus molle l. (pimiento), sus variaciones temporales y su contribución al mejoramiento del inventario de emisiones de la Región Metropolitana. *Chemist Thesis.* Universidad de Chile.

Nilo, C. (2003). Plan Verde: Un instrumento para la gestión y el fomento de áreas verdes en el Gran Santiago. *Urbano*, *6*(8), Universidad del Bío-Bío, Concepción, Chile (in Spanish).

Nowak, D. J. & Sisinni, S. (1993). Plant Chemical Emissions. *Miniature Roseworld*, X, 1, 4-6 *WHO. 1999. Guidelines for air quality, Geneva.*

Pacífico, F., Harrison, S. P., Jones, C. D. & Sitch, S. (2009). Isoprene emissions and climate. *Atmospheric Environment*, *43*, 6121-6135.

Peñuelas J. & Lluisà J. (2002). Linking photorespiration, monoterpenes and thermotolerance in *Quercus. New Phytologist*, *155*, 227-237.

Peñuelas, J. & Staudt, M. (2010). BVOCs and global change. *Trends in Plant Science*, *15*(3), 133-144.

Pichersky, E. & Gershenzon, J. (2002). The formation and function of plant volatiles: perfumes for pollinator attraction and defence. *Current Opinion in Plant Biology*, *5*, 237-243.

Préndez, M. & Peralta, H. (2005). Determinación de Factores de Emisión de Compuestos Orgánicos Volátiles de dos Especies Arbóreas Nativas de la Región Metropolitana, Chile. *Información Tecnológica*, *16*(1), 17-27 (in spanish).

Préndez, M., Alvarado, G. & Serey, I. (2011). Some guidelines to improve air quality management in Santiago, Chile: from commune to basin level. *In: Mazzeo, N. (Ed.), Air Quality Monitoring, Assessment and Management. INTECH Open Access Publisher, 305-328.*

Préndez, M., Carvajal, V., Corada, K., Morales, J., Alarcón, F. & Peralta, H. (2013). Biogenic volatile organic compounds from the urban forest of the Metropolitan Region, Chile. *Environmental pollution*, *183*, 143-150.

Préndez, M., Corada, K. & Morales, J. (2013a). Emission factors of biogenic volatile organic compounds in various stages of growth present in the urban forest of the Metropolitan Region, Chile. *Research Journal of Chemistry and Environment*, *17*(11), 108-116. Published *online* in November 2013 issue.

Prinn, R., Cunnold, D., Rasmussen, R., Simmonds, P., Alyea, F., Crawford, A., Fraser, P. & Rosen, R. (1987). Atmospheric trends in methylcholoroform and the global average for the hydroxyl radical. *Science*, *238*, 945-950.

Puente, M. (2010). Estimación de emisiones biogénicas, Región Metropolitana, año base 2010. *Centro Nacional del medio ambiente CENMA*. Chile. 1-21 (in Spanish).

Reyes, S. & Figueroa, I. (2010). Distribución, superficie y accesibilidad de las áreas verdes en Santiago de Chile. *EURE*, *36*(109), 89-110 (in Spanish).

Sanderson, M. G., Jones, C. D., Collins, W. J., Johnson, C. E. & Derwent, R. (2003). G. Effect of Climate Change on Isoprene Emissions and Surface Ozone Levels, *Geophysical Research Letters*, *30*(18), 1936.

Saunders, S. M., Jenkin, M. E., Derwent, R. G. & Pilling, M. J. (2003). Protocol for the development of the Master Chemical Mechanism, MCM v3 (Part A): tropospheric degradation of non-aromatic volatile organic compounds. *Atmos. Chem. Physic.*, *3*, 161-180.

Schade, G. W., Goldstein, A. H. & Lamanna, M. S. (1999). Are Monoterpene Emissions influenced by Humidity? *Geophys. Res. Letters*, *26*(14), 2187-2190.

Seguel, R., Morales, R. & Leiva, M. (2012). Ozone weekend effect in Santiago, Chile, *Environmental Pollution*, *162*, 72-79.

Seinfeld, J. & Pandis, S. (2006). Atmospheric chemistry and physics. From air pollution to climate change. John Wiley & Sons, 2nd Ed.

Sharkey, T. & Yeh, S. (2001). Isoprene emission from plants. *Plant Physiology and Plant Molecular Biology*, *52*, 407– 436.

Sharkey, T., Singsaas E., Vanderveer, P. & Geron, C. (1996). Field measurements of isoprene emissions from trees in response to temperature and light. *Tree Physiology*, 16, 649-654.

Soja, G. & Soja, A. (1995). Ozone effects on dry matter partitioning and chlorophyll fluorescence during plant development of wheat. *Water, Air and Soil Pollution*, *85*, 1461-1466.

Staudt, M., Bertin, N., Hansen, U., Seufert, G., Ciccioli, P., Foster, P., Frenzel, B., Fugit, J-L & Torres, L. (1997). The BEMA-Project: Seasonal and diurnal patterns of monoterpene emissions from *Pinus pinea* (*L*). Atmos. Environ., *31*, 145-156.

Tausz, M., Grulke, N. E. & Wieser, G. (2007). Defense and avoidance of ozone under global change. *Environmental Pollution*, *147*, 525-531.

Tingey, D. T., Turner, D. P. & Weber, J. A. (1991). Factors controlling the emissions of monotrepenes and other volatile organic compounds. In: *Trace gas emissions by plants*. Eds. T. D Sharkey, E. A. Holland and H. A. Mooney. Academic Press. 93 – 119

Toro, M. V., Cremades, L. & Jairo, J. (2001). Inventario de emisiones biogénicas en el valle de Aburrá, *Revista Ingeniería y Gestión Ambiental*. Universidad Pontificia Bolivariana, *17*, 32-33.

Villanueva-Fierro, I., Popp, C. J. & Martin, R. S. (2004). Biogenic emissions and ambient concentrations of hydrocarbons, carbonyl compounds and organic acids from ponderosa pine and cottonwood trees at rural and forested sites in Central New Mexico, *Atmospheric Environment*, *38*, 249–260.

Went, F. W. (1960). Organic matter in the atmosphere, and its possible relation to petroleum formation. *Proc. Natl. Acad. Sci. USA.*, *46*, 212-221.

WHO, World Health Organization Media Centre, Air Quality and Health, Fact Sheet N°313. 2011. [in line]. http://www.who.int/mediacentre/factsheets/fs313/es/index. html.

Yañez, A. M. (2011). Within-Canopy Sesquiterpene Ozonolysis in Amazonia. Department of Earth and Ecosystem Sciences Physical Geography and Ecosystems Analysis, *Seminar series*, 217, 1-51.

Yokouchi, Y. & Ambe, Y. (1984). Factors affecting the emission of monoterpenos from red pine (*Pinus desinflora*) Long-term effects of light, temperature and humidity. *Plant physiology*, *75*, 1009-1012.

Zimmermann, P. R. (1979). Testing of hydrocarbons emissions from vegetation, leaf liter and aquatic surfaces, and development of a method for compiling biogenic emission inventories. *Rep. EPA-450/4-70-004, U.S. Environmental Protection Agency, Research Triangle Park, N.C.*

In: Volatile Organic Compounds
Editor: Khaled Chetehouna

ISBN: 978-1-63117-862-7
© 2014 Nova Science Publishers, Inc.

Chapter6

LATEST RESULTS ON THE CATALYTIC OXIDATION OF LIGHT ALKANES, AS PROBE VOC MOLECULES, OVER RU-BASED CATALYSTS: EFFECTS OF PHYSICOCHEMICAL PROPERTIES ON THE CATALYTIC PERFORMANCES

Hongjing Wu[1,2], L. F. Liotta[1] and A. Giroir-Fendler[3]*

[1]Istituto per Lo Studio dei Materiali Nanostrutturati (ISMN)-CNR,
Palermo, Italy
[2]Department of Applied Physics,
Northwestern Polytechnical University (NPU), Xi'an, P.R. China
[3]Université Lyon 1, CNRS, UMR 5256, IRCELYON,
Institut de recherches sur la catalyse et l'environnement de Lyon,
Villeurbanne Cedex, France

ABSTRACT

Volatile organic compounds (VOCs) are recognized as having a major responsibility for the increase in global air pollution due to their

---

* Corresponding Author address. Email: liotta@pa.ismn.cnr.it; Tel.: +39-091-6809-371; Fax: +39-091-6809-399.

contribution to ozone and photochemical smog. Currently, the most active catalysts for VOCs oxidation are based on noble and transition metals.

Among noble metals, platinum, palladium and rhodium exhibit high activity and selectivity at low temperature, but they are unstable in the presence of chloride compounds.

Systems based on transition metal oxides (such as V_2O_5, MnO_2, Co_3O_4-based oxides) are generally less active than noble metals, some formulations are stable to chlorines and their cost is significantly lower.

Supported ruthenium catalysts have received much attention over the past years, because of their high activity in oxidation as well as reduction reactions. Such catalysts have proved to be among the best catalytic systems for oxidation of various substrates, such as carbon monoxide, ammonia, hydrogen, alcohols, diesel soot, and even in low temperature oxidation of HCl. However, few studies on VOCs oxidation over supported Ru have been conducted so far, especially in light alkane combustion. The physicochemical properties of Ru catalysts have been found to strongly influence activity and stability. Therefore, the comparison between catalysts prepared by different methods and pre-treated in different conditions is difficult.

In order to give a general overview on the state of art, the present chapter focuses on the latest results on the catalytic performance of Ru catalysts for light alkanes oxidation with special attention to the structure-activity relationship.

Keywords: Ru, oxide, catalytic oxidation, light alkanes, structure sensitivity

LIST OF SYMBOLS AND ACRONYMS

VOCs	volatile organic compounds
Pt	platinum
Pd	palladium
Rh	rhodium
Ru	ruthenium
EG	ethylene glycol
PVP	poly(N-vinyl-2-pyrrolidone)
MW	microwave
HRTEM	high-resolution transmission electron microscopy
TEM	transmission electron microscopy
RT	room temperature
XPS	x-ray photoelectron spectroscopy

SAED	selected area electron diffraction
SMSI	strong metal-support interaction
FFT	fast Fourier transform
TPR	temperature programmed reduction

INTRODUCTION

Volatile organic compounds (VOCs) are recognized as having a major responsibility for the increase in global air pollution due to their contribution to ozone and photochemical smog [1]. The nature of VOCs depends on the source of emissions and comprises a large variety of compounds, such as alkanes, olefins, alcohols, ketones, aldehydes, aromatics, and halogenated hydrocarbons [2]. Alkanes are the major families of pollutants in industrial emissions. Among the methods for reducing emissions of VOCs from major stationary sources, catalytic combustion is the most effective for low concentrations of organic emissions [3, 4]. Selection of the catalysts for total oxidation various organic pollutants have been the subject of many studies. It is not an easy task.

Two classes of catalysts, noble metals and transition metal oxides, are the most promising catalysts in the area of VOCs combustion [5, 6]. Generally, among the supported metal catalysts, platinum (Pt), palladium (Pd), rhodium (Rh) and gold (Au) have been studied the most, but other metals, such as ruthenium (Ru), were also reported as active catalysts for alkane oxidation.

On the other hand, ruthenium has been much less investigated, despite the fascinating chemistry of this noble metal and its much lower cost (10 or 20 times less expensive than Pt and Pd, respectively) [7]. Ruthenium is implemented in electrochemistry as supercapacitors [8, 9], electrodes [10], or electrocatalysts [7]. As a catalyst, it is known to be very active in the hydrogenation of CO_2 [11-13], in the oxidation of CO [14, 15], in the Deacon process [16, 17], in ammonia synthesis [18], and methanation of CO_2 at low temperature [11]. Recently, Okal's group has reported a series of studies, which clearly demonstrate the high potential of supported Ru catalysts for abatement of VOCs, especially for light alkanes [19-22].

Our present chapter is devoted to the application of Ru catalysts for the total oxidation of light alkanes, as probe VOC molecules. Until now, none of the recent reviews or books dealing with catalysts for VOC oxidation have been devoted to such catalytic systems. We considered it worth focusing on the latest catalytic results with special attention to the structure-activity relationship and catalysts stability.

RESULTS AND DISCUSSION

General

Catalytic oxidation of light alkanes, alkenes and aromatics have been widely investigated over metal catalysts such as Pt, Pd and Rh, and many studies have shown that short-chain hydrocarbons are amongst the most difficult to degrade [23, 24]. It is well known that both the noble metal and the oxide support play an essential role in the catalytic oxidation [25]. For this reason, a number of recent studies have focused on new catalytic oxidation catalysts based on noble metals, such as Pt, Pd, Rh and Au, supported on various metal oxides or mixed oxide supports such as Co_3O_4-CeO_2 [26, 27]. Supported Ru catalysts have also attracted much interest over the past years, because of their high oxidation activity at low temperature [28-33]. The catalytic activity for oxidation of ethyl acetate, acetaldehyde, and toluene was investigated over Ru based catalysts deposited on supports such as γ-Al_2O_3, CeO_2, ZrO_2 and SnO_2 [30]. Recently, reactivity of Ru supported on CeO_2 and γ-Al_2O_3 in the oxidation of propene, toluene and carbon black has been studied [33]. Moreover, the performance of Ru supported on γ-Al_2O_3 catalysts for the oxidation of n-/iso-butane mixture and propane has been investigated [19, 34]. Generally, amongst the supports examined in literature, the CeO_2 oxide is one of the most promising candidates for oxidation of light alkanes quite likely due to the good redox properties and its crucial role as a source of oxygen. Rare metal oxide such as ceria also promotes stabilization of noble metals and prevents sintering of nanoparticles [35].

Ru/γ-Al_2O_3

The study of metal nanoparticles on oxide supports is of importance in catalysis because these systems play a central role in industrial processes and in pollution abatement. The mean particle size and particle size distribution of the metal nanoparticles is often critical in determining catalytic activity and selectivity [36]. In recent years, well-defined supported Ru nanoparticles were successfully synthesized via controlled colloid chemistry routes. Ru nanoparticles have been obtained by reduction of $RuCl_3$ in glycols, such as ethylene glycol (EG) in the presence of the inorganic support, hydrogen reduction or sodium borohydride reduction. Recently, colloidal Ru/Al_2O_3 catalysts have been prepared by the reduction of $RuCl_3$ in a low boiling point alcoholic solution under overpressure [37]. In order to stabilize the colloid nanoparticles from agglomeration, metal salts are reduced in the presence of organic compounds as stabilizers. However, the protected Ru nanoparticles are not stable under treatment at higher temperature since such treatment leads to a sintering of particles.

Supported Ru nanoparticles are prepared without using any stabilizing agents. To prevent their aggregation, capping agents like poly(N-vinyl-2-pyrrolidone) (PVP) and dendrimers are used. The removal of PVP from the $Ru(PVP)/\gamma-Al_2O_3$ catalyst is difficult but necessary to obtain active catalytic systems. Moreover, the removal of ligands or the templating agents also leads to metal sintering. Although a growing number of contributions deal with supported Ru catalysts, not much information is available concerning the resistance towards sintering of such systems at relatively high temperatures.

Okal et al. have focused on the characterization and thermal stability of the colloidal low-loaded $Ru/\gamma-Al_2O_3$ catalyst obtained by microwave (MW)-assisted synthesis [38]. The catalyst was prepared without using any stabilizing agent. The thermal stability of the supported Ru colloid was evaluated by heating the as-prepared catalyst in hydrogen at 500-700°C. Such studies are important since Ru^0 nanoparticles are expected to have a weaker interaction with the support than in the catalysts obtained by the impregnation method [39]. Such weaker interaction with the support leads to the easier agglomeration of the Ru^0 nanoparticles. Their results showed the dispersion of the fresh catalyst decreased slightly from 0.53 to 0.49 by hydrogen treatment at 700°C, and good correspondence with high resolution transmission electron microscopy (HRTEM) data was obtained [38]. Results indicated that some migration and coalescence of small Ru nanoparticles occurred at 700°C, and high stability of the Ru nanoparticles in the low-loaded colloidal $Ru/\gamma-Al_2O_3$ catalyst under reducing conditions could be obtained.

Furthermore, Zawadzki and Okal have reported that uniform and stable Ru colloids were synthesized by reduction of $RuCl_3$ in ethylene glycol (EG) in the presence of PVP by using a microwave-assisted solvothermal method [40]. A microwave accelerated reaction system MW Reactor Model 02-02 (ERTEC, Poland) was used to carry out the microwave-assisted solvothermal reactions. In detail, 40 mg $RuCl_3$ was first dissolved in 150 ml of EG, next 0.85 g of PVP was added under magnetic stirring at room temperature (RT), and after the solids had dissolved completely, the 50 ml mixture was put in the reaction vessel.

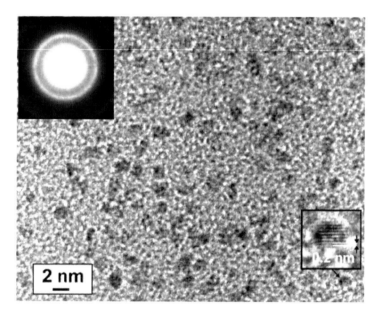

Figure 1. The typical TEM image of the PVP-coated Ru colloids obtained under microwave-assisted solvothermal method [40].

The microwave-solvothermal reactions were carried out under mild conditions of temperature, 180 (colloid 1) or 250°C (colloid 2), with a hold time of 10 or 30 min, respectively. The Ru colloid was supported on γ-Al_2O_3 by two different methods: (catalyst 1) immobilization of the PVP-stabilized Ru colloid, which was already formed, onto alumina carrier and (catalyst 2) in situ deposition, e.g., reduction of $RuCl_3$ with EG in the presence of the support, which prevented the agglomeration of the Ru colloids.

Figure 1 shows a representative TEM image of the PVP-stabilized Ru nanoparticles prepared using microwave synthesis. It appears the Ru colloids are well separated with no agglomeration tendency. At higher magnification, it is seen that the colloids are single crystals exhibiting lattice fringes characteristic for Ru^0. The electron diffraction pattern of the representative Ru colloid is also shown in Figure 1 (inset); it exhibits broad rings with the *d*-spacings of *hcp* Ru metal. It suggests that the colloids are small and/or are of low crystallinity. The particle size distributions obtained from TEM images (not shown) are fairly narrow and an average particle size is estimated to be ~ 1.7 and 1.9 nm for the colloid 1 and 2, respectively.

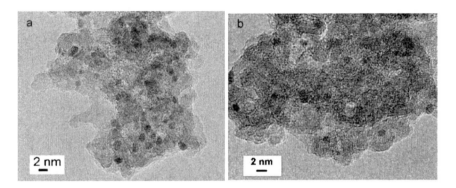

Figure 2. The typical TEM images of supported colloidal Ru catalyst 1 (a) and 2 (b) obtained by using different methods [40].

Representative TEM images of the supported Ru catalysts 1 and 2 (Figure 2a and 2b) show the presence of numerous dark and well-dispersed particles on the support. HRTEM images (not reported) show that these spots correspond to the Ru particles. It can be seen that individual Ru particles have no clear edge and are similarly quasi-spherical and quite uniform in size. The average diameter of Ru particles estimated from Gaussian fit of the particle size distribution histograms is ~ 1.8 and 2.1 nm, respectively, and corresponds well to the average size of the unsupported PVP-stabilized Ru colloids (see Figure 1). X-ray photoelectron spectroscopy (XPS) indicates that Cl^- from Ru precursor are partially incorporated into the system obtained from pre-prepared PVP-stabilized Ru colloids, while in situ deposition results in the Cl^- free system. The PVP free catalyst shows excellent chemisorption properties that are assigned to the high dispersion of the small Ru colloids and also to their weak interaction with the support. The stabilizer, such as PVP, covers the metal surfaces and leads to the low metal surface area. Thus, this work demonstrates that the removal of PVP from the catalyst is necessary to obtain an active catalytic system.

It is known that the activity and selectivity of the Ru catalysts depend strongly on the oxidation state of the metal since different Ru oxides exhibit distinct physical and catalytic properties. For this reason, the interaction between Ru and oxygen has recently been a subject of intensive investigations [41]. The interaction of oxygen with alumina-supported Ru at room temperature was studied thoroughly in connection with the problem of determination of the metal surface area and considerably less data refers to the metal oxidation at higher temperatures [42]. Therefore, the interaction of oxygen with the high-loaded Ru/γ-Al$_2$O$_3$ catalyst prepared by incipient wetness from RuCl$_3$ over a wide temperature range, 20-400°C, was reported by Okal [43]. The effect of residual Cl$^-$ precursors on the Ru oxidation was also investigated. It is generally known that Cl$^-$ ions affect both catalytic and chemisorptive properties of heterogeneous catalysts. The results here presented reveal that Cl$^-$ ions affect not only chemisorptive but also oxidative properties of supported Ru nanoparticles. At low temperatures (100-200°C), very small particles are completely oxidized forming amorphous Ru oxide. At 250°C, formation of poorly ordered RuO$_2$ with mean crystallite size of ~ 4 nm is found, but large Ru particles are covered with a very thin nearly amorphous oxide layer. At 300 and 400°C, the crystallization of the RuO$_2$ phase and significant sintering of oxide particles is observed (see Figure 3). HRTEM and SAED data confirm the presence of Ru^0 even at 400°C. Large Ru particles are covered with ~ 5 nm layer of crystalline RuO$_2$, while small Ru particles are completely transformed into RuO$_2$ phase. It means that oxidation of supported Ru depends not only on the oxidation temperature but also on metal particle sizes. The results are useful in understanding the processes occurring during the activation step of the Ru catalysts used for catalytic oxidation of VOCs.

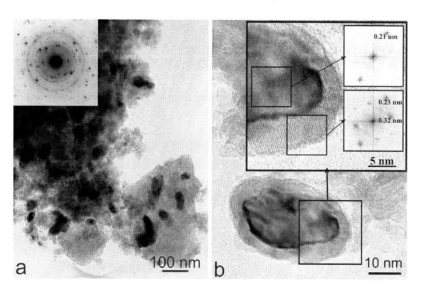

Figure 3. TEM images (a) and SAED pattern (b) of the washed 10.8% Ru/γ-Al$_2$O$_3$ catalyst after oxidation at 400°C [43].

Okal et al. have studied the effect of the Ru/γ-Al$_2$O$_3$ catalysts (prepared by an incipient wetness impregnation method) pre-treatment (reduction in H$_2$ or oxidation in O$_2$), and the effect of residual chlorines on propane oxidation [34]. It was established that in the presence of a large amount of Cl$^-$ ions on the catalyst surface, activities of the catalysts towards the propane oxidation reaction were greatly suppressed. In the case of the fresh reduced Ru/γ-Al$_2$O$_3$ catalysts the initial activity in propane oxidation was related to the presence of oxygen-deficient Ru$_x$O$_y$ particles or to a few thick layers of surface oxide on the large Ru particles (Figure 4). In these small particles rather weakly-bond chemisorbed species are expected. Oxidation of the Ru/γ-Al$_2$O$_3$ catalysts at 250°C, was connected with the formation of RuO$_2$ crystallites (~ 4 nm) and the oxygen-deficient oxide layer on the large Ru particles. Stronger Ru-O bonds in these small crystallites caused some decrease in the propane oxidation activity. Oxidation at 600°C, induced significant loss activity of the Ru/γ-Al$_2$O$_3$ catalysts. It could be ascribed to the strong Ru-O bonds in the large RuO$_2$ crystallites, as well as to the decreasing number of active sites caused by significant sintering of the Ru phase. Our data indicate that the Ru/γ-Al$_2$O$_3$ catalysts, without Cl$^-$, may be suitable as catalysts for the low temperature oxidation of a low concentration of propane.

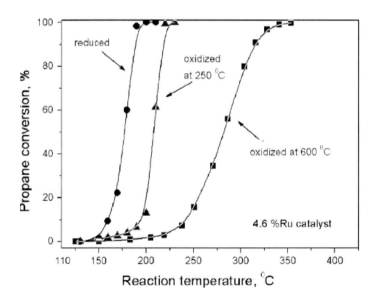

Figure 4. Propane conversion over the washed 4.6 wt.% Ru/γ-Al$_2$O$_3$ catalyst after various pre-treatments [34].

Table 1. Temperatures (°C) required for 50% and 95% conversion of propane and TOF at 170°C on various catalysts

Catalysts	Light-off curves		TOF (s^{-1})
	$T_{50\%}$ (°C)	$T_{95\%}$ (°C)	
As-prepared			
First cycle	176	189	
Second cycle	185	197	0.94×10^{-3}
Hydrogen treated			
First cycle	186	197	
Second cycle	194	203	0.41×10^{-3}

Furthermore, Ru/γ-Al$_2$O$_3$ catalysts have been prepared by a one-step microwave-polyol method and also tested in the complete oxidation of propane [22]. It was found that the propane oxidation turnover rate on as-prepared catalyst with mean Ru particle size of 1.6 nm and narrow size distribution was about two times higher than the same after hydrogen pre-treatment, which contained Ru nanoparticles with a mean size of 6 nm (see Table 1). It was revealed that except for the high metallic dispersion and small Ru nanoparticle sizes, the co-existence of metallic and oxide Ru species played an important role for propane oxidation, but the most active sites seemed to consist of small Ru$_x$O$_y$ clusters and a poorly ordered layer of RuO$_2$ on the large Ru crystallites. Formation of such surface species in both catalysts at 100-200°C allowed them to reach 100% propane conversion at temperatures below 210°C. Moreover, under oxidative conditions up to 250°C, both in oxygen as well as during the reaction of propane oxidation, Ru/γ-Al$_2$O$_3$ are characterized by good stability.

Okal et al. have also investigated the catalytic performance of Ru/γ-Al$_2$O$_3$ catalysts prepared by the incipient wetness method in oxidation of n-/iso-butane [19]. It was found that the pre-treatment has considerable influence on activity of the Ru/γ-Al$_2$O$_3$ catalysts. The severe contamination of Ru particles by Cl$^-$ results in lowering the activity of the Ru/γ-Al$_2$O$_3$ catalysts. It was established that reduced Cl$^-$ free catalysts exhibited higher activity than the ones oxidized at 250°C. Catalytic and characterization results revealed that the most active sites in the butane oxidation, probably consist of a few layers of thick surface oxide on the large Ru particles and small Ru$_x$O$_y$ clusters without well-defined stoichiometry. Such active surface species are formed mainly in the most active catalyst, the reduced 4.6 wt.% Ru/γ-Al$_2$O$_3$, which reaches 100% butane conversion below 200°C (see Figure 5).

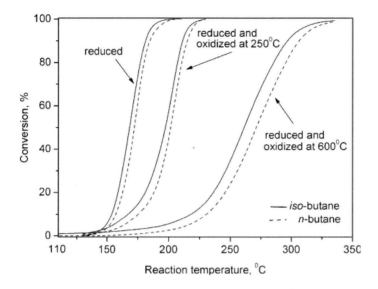

Figure 5. Butane conversion over the 4.6 wt.% Ru/γ-Al$_2$O$_3$ catalyst (washed) after oxidation pre-treatments at 250 and 600°C as a function of reaction temperature [19].

The catalytic performance of Ru declines as the catalysts are oxidized at a higher temperature. The activity loss is attributed to the formation of the crystalline RuO$_2$ phase and to some sintering of the active phase. In the used catalysts, small Ru particles are oxidized to RuO$_2$ while the large Ru particles are covered with a RuO$_2$ layer, with a thickness of 2-3 nm.

Ru/ZnAl$_2$O$_4$

Metal aluminates with spinel structure, among others ZnAl$_2$O$_4$, are of interest due to their technological applications as refractory, optical and catalytic materials [44]. More importantly, nanostructured and porous aluminates with high specific surface area are of special interest due to their improved properties such as lower temperature sinter-ability, greater thermal and chemical stability, increased hardness, better diffusion, etc. Therefore, in recent years, nanocrystalline ZnAl$_2$O$_4$ spinels have attracted considerable interest in the field of the catalysts since they may be applied as supports for noble metal to substitute traditional supports such as γ-Al$_2$O$_3$. As a result, ZnAl$_2$O$_4$ spinel has been proposed as an excellent palladium support for hydrogenation reactions, as well as for platinum or bimetallic Pt-Sn catalysts used in the dehydrogenation of alkanes. Zinc alumina spinel may also be a good candidate for catalytic materials for various reactions, such as VOC oxidation and soot oxidation. Moreover, it was found that zinc alumina spinel could also exhibit the strong metal-support interaction (SMSI) with noble metals [45]. So far, there is only scarce data concerning the application of the zinc alumina spinel as support material in VOCs oxidation.

Zinc alumina, with spinel structure and high specific surface area, was never used as support for Ru catalysts. The structure and activity of Ru catalysts supported on $ZnAl_2O_4$ and CeO_2 is compared to that supported γ-Al_2O_3 [21]. The $ZnAl_2O_4$ spinel was prepared by the unconventional co-precipitation method using aqueous solutions of $Zn(NO_3)_2$ and $Al(NO_3)_3$. The precipitate was filtered off, washed with water, then air-dried and finally calcined at 550°C for 3 h. The catalysts with Ru loading of ~ 5 wt.% were prepared by the incipient wetness impregnation method using $Ru(NO)(NO_3)_3$ as a metal precursor. The impregnated materials were air-dried at 120°C for 20 h and finally reduced in H_2 flow at 500 or 400°C for 5 h.

Typical TEM images and SAED patterns (insets) for all catalysts are shown in Figure 6. For the $Ru/ZnAl_2O_4$ and Ru/CeO_2 catalysts, metal particles could be identified only by HRTEM images (right) and the fast Fourier transform (FFT) pattern (inset). These results indicate good dispersion of Ru particles. Only in the SAED pattern of the Ru/γ-Al_2O_3 catalyst weak diffraction spots from the Ru metal phase are present. The mean particle size of 1.7 nm and 1.5 nm is obtained for the $Ru/ZnAl_2O_4$ and Ru/CeO_2 catalysts; while for the Ru/γ-Al_2O_3 catalyst a much broader distribution of particle sizes is observed (1-9 nm) with a mean size of 2.6 nm.

Okal et al. have further prepared highly dispersed $Ru/ZnAl_2O_4$ catalysts by changing the Ru loading from 0.5 to 4.5 wt.% [20]. Good activity of the 0.5-4.5 wt.% $Ru/ZnAl_2O_4$ catalysts was attributed to the high dispersion of the Ru species present in these materials. Dispersion of the $Ru/ZnAl_2O_4$ catalysts was high and decreased from 71 to 56% with the rise of Ru content from 0.5 to 4.5 wt.%. The O_2 uptake results showed that Ru in the 0.5-1 wt.% $Ru/ZnAl_2O_4$ catalysts was oxidized already at 150°C, while in the 4.5 wt.% catalyst at 400 °C. The catalyst activity correlated well with a high dispersion and small Ru particle sizes. Under oxidative conditions up to 250°C, both in oxygen as well as during reaction of propane oxidation, Ru clusters supported on $ZnAl_2O_4$ spinel, especially at low-metal loading, are characterized by very good stability (see Figure 7). It was established that activity of the regenerated hydrogen $Ru/ZnAl_2O_4$ catalysts (with larger Ru particles) is lower as compared to the fresh reduced samples.

Latest Results on the Catalytic Oxidation of Light Alkanes ...   155

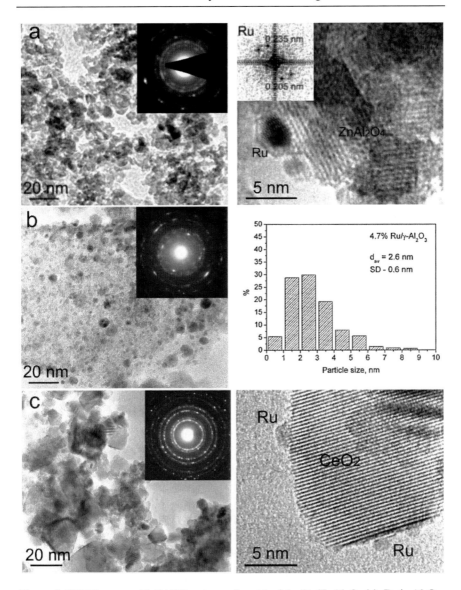

Figure 6. TEM images with SAED patterns (insets) of the Ru/ZnAl$_2$O$_4$ (a), Ru/γ-Al$_2$O$_3$ (b), and Ru/CeO$_2$ (c) catalysts [21].

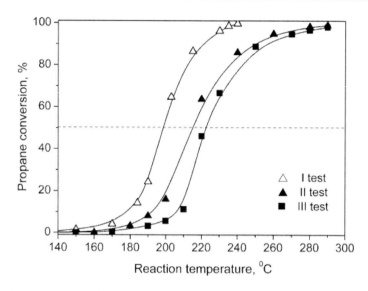

Figure 7. Propane conversion over the 4.5 wt.% Ru/ZnAl$_2$O$_4$ catalyst after regeneration [20].

Recently, the activity of the Ru/ZnAl$_2$O$_4$ catalysts, H$_2$-reduced or air-aged at 700°C, for the catalytic oxidation of methane under O$_2$-rich conditions has been studied [46]. The size of Ru particles in fresh catalysts varied from 1.1 to 1.5 nm with the metal loading from 0.5 to 4.5 wt.%. Air aging treatment caused severe agglomeration of the Ru phase and formation of the well-crystallized RuO$_2$ oxide. Under reaction conditions, highly dispersed Ru species were easily oxidized and RuO$_2$ was the active phase for methane oxidation. The fresh catalysts are more effective than aged samples in terms of light-off temperature and temperature needed for the complete methane conversion (see Figure 8). The mean crystallite size of the RuO$_2$, formed during combustion reaction over fresh catalysts, depended on metal loading and was lower (21-27 nm) as compared to that formed during the aging process (21-40 nm), which leads to higher activity. However, stable catalytic activity was observed only for aged catalysts. The catalytic activity could be partly explained by the changes in the morphology and the crystallite size of the RuO$_2$ phase and may suggest a structure sensitivity of CH$_4$ combustion over Ru catalysts.

HRTEM images and SAED patterns (insets) of the fresh 1 and 4.5 wt.% Ru catalysts before and after methane combustion reaction (shown in Figure 9 and 10, respectively) indicate drastic changes in the morphology of the Ru phase. After the catalytic tests, the HRTEM images and SAED patterns (insets) confirm the presence of crystalline RuO_2 particles with lattice fringes of distance 0.317 nm and contain strong reflections, which could be assigned to the RuO_2 oxide structure. The HRTEM image in Figure 10b shows that the morphology of the RuO_2 crystallites is rod-like with large particle dimensions of about 15 nm × 60 nm. Comparison of Figure 9b and 10b reveals that RuO_2 crystallites are smaller and more irregular for the used 1 wt.% Ru catalyst (particle dimensions of about 14 nm × 28 nm) than those observed in the used 4.5 wt.% Ru catalyst. The different morphologies of the Ru oxide particles in the deactivated catalysts with low and high Ru loading, clearly suggest that a strong structure sensitivity effect may occur for combustion of methane on Ru oxide.

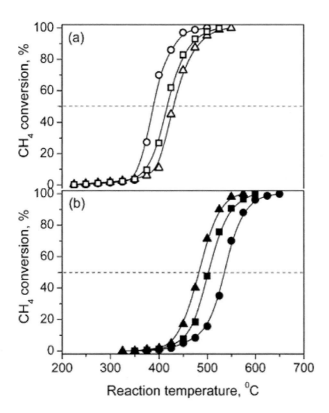

Figure 8. Methane conversion over the fresh (a) and aged (b) $Ru/ZnAl_2O_4$ catalysts with Ru loading of 0.5 (Δ, ▲), 1 (□, ■) and 4.5 (○, ●) wt. % as a function of temperature [46].

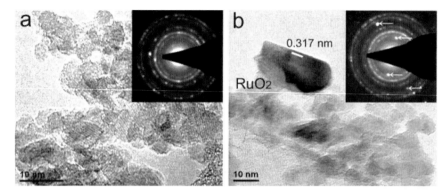

Figure 9. HRTEM images and SAED patterns (insets) of the 1 wt.% Ru/ZnAl$_2$O$_4$ catalyst reduced at 500°C (a) and after catalytic tests in methane combustion (b) [46].

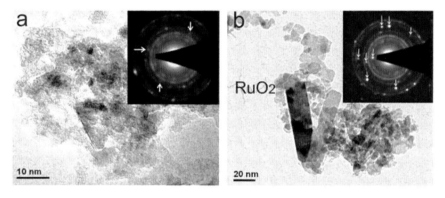

Figure 10. HRTEM images and SAED patterns (insets) of the 4.5 wt.% Ru/ZnAl$_2$O$_4$ catalyst reduced at 500°C (a) and after catalytic tests in methane combustion (b) [46].

Highly dispersed Ru/ZnAl$_2$O$_4$ catalysts (0.5 and 4.5 wt.% Ru) have been further treated in air at 300-600°C and their catalytic performance for the methane combustion were investigated [47]. Catalytic data showed that pre-treatment procedure had a significant influence on the activity of the Ru catalysts. The catalysts treated at 300°C are more active than the calcined one in the air at 400°C and those treated at higher temperatures (see Figure 11). The increase in Ru loading from 0.5 to 4.5 wt.% caused a decrease of T_{50} by ~ 100°C. The best catalytic results were obtained over the high-loaded Ru catalyst treated at 300-400°C, which was able to completely convert CH$_4$ to CO$_2$ at temperature ~ 510°C. TEM data shows that small Ru nanoparticles (~ 1.5 nm) undergo oxidation during thermal treatment in air with the formation of larger RuO$_2$ particles (see Figure 12), which leads to a significant decrease of the Ru dispersion. For the 0.5 and 4.5 wt.% Ru catalysts treated at 300-600 °C, the mean RuO$_2$ crystallite size increases from 18 to 31 nm and 16 to 33 nm, respectively. The large change of the crystallite size of the Ru oxide, in the catalyst at the same Ru content, causes the exposed RuO$_2$ surface to reduce and this leads to a decrease of the specific reaction rate per gram of Ru. However, the activity of Ru/ZnAl$_2$O$_4$ catalysts was very stable when it was treated at 500°C and above.

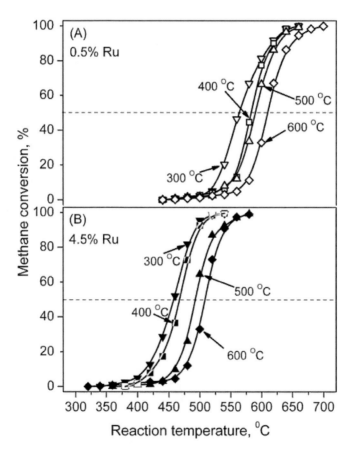

Figure 11. Methane conversion over the air treated at 300-600°C Ru/ZnAl$_2$O$_4$ catalysts with the Ru loading of 0.5 (A) and 4.5 wt.% (B) as a function of temperature [47].

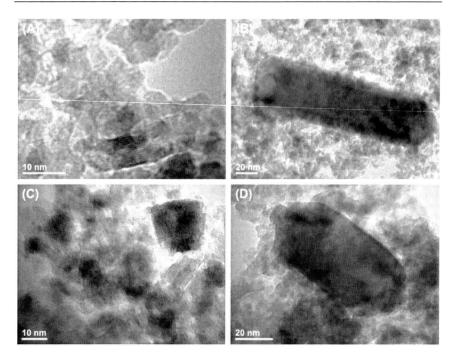

Figure 12. HRTEM images of the 0.5 (A and B) and 4.5 wt.% (C and D) Ru/ZnAl$_2$O$_4$ catalyst reduced at 500°C and next treated in air for 3 h at 300°C (A and C) and at 500 °C (B and D) [47].

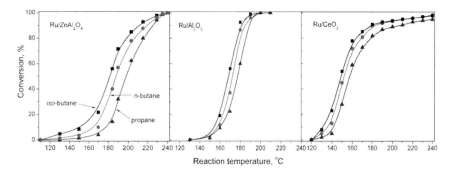

Figure 13. Conversion of iso-butane (■), n-butane (●), and propane (▲) over supported Ru catalysts as a function of temperature [21].

Ru/CeO$_2$

A few studies have been devoted to the total oxidation of VOCs over a Ru/CeO$_2$ catalyst. However, literature data showed that Ru oxide catalysts are active in acetic acid [48] and propene oxidation reactions [28]. The pronounced reactivity of Ru oxide catalysts seemed to result from the reducibility of the oxide itself. In the presence of another reducible oxide such as ceria (CeO$_2$), the oxygen needed for the oxidation reaction seemed to be provided by the second oxide.

It was shown that the addition of Ru to the different supports was very beneficial for the total oxidation of propene [29]. Temperature programmed reduction (TPR) experiments of the catalysts showed that the oxygen species of Ru oxides were reduced at low temperatures, which was the main reason for its high reactivity in oxidation reactions. Eguchi et al. reported that a remarkable effect of the support was observed, and Ru/CeO$_2$ exhibited the highest activity for all tests regardless of the reduction treatment in H$_2$ atmosphere [30]. The redox species of Ru on CeO$_2$ with high dispersion should be responsible for the activity, which will be the useful design guide for low-temperature VOCs oxidation. On the other hand, the catalytic activity of Ru/ZrO$_2$ and Ru/γ-Al$_2$O$_3$ was enhanced by the reduction treatment due to the formation of Ru^0. In the case of Ru/SnO$_2$, the formation of intermetallic compound with core-shell structure was confirmed, resulting in the deterioration of catalytic performance. These results confirm that the effect of pre-treatments on catalytic activity depends on the kinds of supports used for the Ru metal. They further studied the noble metal effect on the catalytic activity for ethyl acetate combustion and found that Ru/CeO$_2$ achieved the highest activity [32]. It was confirmed by TPR that the reduction of the Ru species was initiated at the lowest temperature among the CeO$_2$-supported precious metals (Pt, Pd and Rh). The precious metal species, reducible at lower temperatures, should be responsible for the high activity in the complete oxidation of ethyl acetate. Finally, Okal et al. recently reported that the Ru/CeO$_2$ catalyst was tested in the total oxidation of propane and n-/iso-butane [21]. All oxidation reactions occur at much lower temperatures over Ru/CeO$_2$ compared to those over Ru/γ-Al$_2$O$_3$ and Ru/ZnAl$_2$O$_4$ catalysts (see Figure 13). The redox species of Ru on the CeO$_2$ oxide, easily reacted with the lattice oxygen of CeO$_2$, are responsible for the enhanced activity of the Ru/CeO$_2$ catalyst in the VOCs oxidation.

CONCLUSION

In this chapter we have discussed the influence of the metal loading, thermal treatment in reducing or oxidizing atmosphere (hydrogen or air), and the calcination procedure on the structure and catalytic properties of Ru catalysts for the light alkanes total oxidation. The results so far discussed lead to the following conclusions:

In the presence of a large amount of Cl^- ions on the catalyst surface, activities of the catalysts towards alkanes oxidation are greatly suppressed.

Except for the high metallic dispersion and small Ru nanoparticle sizes, the co-existence of metallic and oxide Ru species play an important role for light alkanes oxidation, but the most active sites seem to consist of small Ru_xO_y clusters and a poorly ordered layer of RuO_2 grown on the large Ru crystallites.

All oxidation reactions occur at much lower temperatures over Ru/CeO_2 compared to those over $Ru/\gamma-Al_2O_3$ and $Ru/ZnAl_2O_4$ catalysts. The redox species of Ru on CeO_2 oxide easily react with the lattice oxygen of CeO_2, and are responsible for the enhanced activity of the Ru/CeO_2 catalyst in the light alkanes oxidation.

The pretreatment temperature and atmosphere (reducing or oxidizing) has considerable influence on activity of the Ru catalysts. The catalytic performance of Ru declines as the catalysts are oxidized at high temperature, such as 600°C. A large change of the crystallite size of the RuO_2 oxide, in catalysts with the same Ru content, causes a decrease of the exposed RuO_2 surface with a consequent decrease of the specific reaction rate per gram of Ru, suggesting a structure sensitivity of light alkanes combustion over Ru catalysts.

Based on this, in our opinion, not enough attention has been devoted to the durability of Ru-based catalysts under the reaction conditions. Moreover, taking into account the high exothermicity of the VOC oxidation and the strong effect of the pretreatment temperature and atmosphere on the catalytic performances, the poor stability of Ru catalysts appears to be the major drawback for any commercial application.

ACKNOWLEDGMENTS

The authors acknowledge the financial support provided by COST Action CM 1104, as well the financial support provided by the China Scholarship Council for supporting Mr. Wu's scholarship. H. Wu thanks the Excellent Doctorate Foundation, the Doctorate Foundation of Northwestern Polytechnical University, and the Scholarship Award for Excellent Doctoral Student granted by the Ministry of Education.

REFERENCES

[1] S. Scirè, L. F. Liotta, Supported gold catalsyts for the total oxidation of volatile organic compounds, *Applied Catalysis B*, vol. 125, pp. 222-246, 2012.

[2] R. M. Heck, R. J. Farrauto, Catalytic pollution control, second ed., Wiley Interscience, New York, 2002.

[3] S. Ojala, S. Pitkaaho, T. Laitinen, N. Niskala Koivikko, R. Brahmi, J. Gaalova, L. Matejova, A. Kucherov, S. Paivarinta, C. Hirschmann, T. Nevanpera, M. Riihimaki, M. Pirila, R. L. Keiski, Catalysis in VOC abatement, *Topics in Catalysis*, vol. 54, pp. 1224-1256, 2011.

[4] J. J. Spivey, Complete catalytic oxidation of volatile organics, *Industrial & Engineering Chemistry Research*, vol. 26, pp. 2165-2180, 1987.

[5] W. B. Li, J. X. Wang, H. Gong, Catalytic combustion of VOCs on non-noble metal catalysts, *Catalysis Today*, vol. 148, pp. 81-87, 2009.

[6] L. F. Liotta, Catalytic oxidation of volatile organic compounds on supported noble metals, *Applied Catalysis B*, vol. 100, pp. 403-412, 2010.

[7] H. Over, Surface chemistry of ruthenium dioxide in heterogeneous catalysis and electrocatalysis: from fundamental to applied research, *Chemical Review*, vol. 112, pp. 3356-3426, 2012.

[8] C. Sassoye, C. Laberty, H. Le Khanh, S. Cassaignon, C. Boissière, M. Antonietti, C. Sanchez, Block-copolymer-templated synthesis of electroactive RuO_2-based mesoporous thin films, *Advanced Functional Materials*, vol. 19, pp. 1922-1929, 2009.

[9] J. P. Zheng, T. R. Jow, A new charge storage mechanism for electrochemical capacitors, *Journal of the Electrochemical Society*, vol. 142, pp. L6-L8, 1995.

[10] O. Delmer, P. Balaya, L. Kienle, J. Maier, Enhanced potential of amorphous electrode materials: Case study of RuO_2, *Advanced Materials*, vol. 20, pp. 501-505, 2008.

[11] C. Sassoye, G. Muller, D. P. Debecker, A. Karelovic, S. Cassaignon, C. Pizarro, P. Ruiz, C. Sanchez, A sustainable aqueous route to highly stable suspensions of monodispersed nano ruthenia, *Green Chemistry*, vol. 13, pp. 3230-3237, 2011.

[12] C. Hao, S. Wang, M. Li, L. Kang, X. Ma, Hydrogenation of CO_2 to formic acid on supported ruthenium catalysts, *Catalysis Today*, vol. 160, pp. 184-190, 2011.

[13] M. Kuśmierz, Kinetic study on carbon dioxide hydrogenation over Ru/γ-Al_2O_3 catalysts, *Catalysis Today*, vol. 137, pp. 429-432, 2008.

[14] D. W. Goodman, C. H. F. Peden, M. S. Chen, CO oxidation on ruthenium: The nature of the active catalytic surface, *Surface Science*, vol. 601, pp. L124-L126, 2007.

[15] S. H. Joo, J. Y. Park, J. R. Renzas, D. R. Butcher, W. Huang, G. A. Somorjai, Size effect of ruthenium nanoparticles in catalytic carbon monoxide oxidation, *Nano Letters*, vol. 10, pp. 2709-2713, 2010.

[16] D. Crihan, M. Knapp, S. Zweidinger, E. Lundgren, C. J. Weststrate, J. N. Andersen, A. P. Seitsonen, H. Over, Stable deacon process for HCl oxidation over RuO_2, *Angewandte Chemie International Edition*, vol. 47, pp. 2131-2134, 2008.

[17] D. Teschner, R. Farra, L. Yao, R. Schlögl, H. Soerijanto, R. Schomäcker, T. Schmidt, L. Szentmiklósi, A. P. Amrute, C. Mondelli, J. Pérez-Ramírez, G. Novell-Leruth, N. López, An integrated approach to Deacon chemistry on RuO_2-based catalysts, *Journal of Catalysis*, vol. 285, pp. 273-284, 2012.

[18] C. Fernández, C. Sassoye, D. P. Debecker, C. Sanchez, P. Ruiz, Effect of the size and distribution of supported Ru nanoparticles on their activity in ammonia synthesis under mild reaction conditions, *Applied Catalysis A*, doi: 10.1016/j.apcata.2013.1009.1039, 2013.

[19] J. Okal, M. Zawadzki, Catalytic combustion of butane on Ru/γ-Al_2O_3 catalysts, *Applied Catalysis B*, vol. 89, pp. 22-32, 2009.

[20] J. Okal, M. Zawadzki, Combustion of propane over novel zinc aluminate-supported ruthenium catalysts, *Applied Catalysis B*, vol. 105, pp. 182-190, 2011.

[21] J. Okal, M. Zawadzki, L. Krajczyk, Light alkane oxidation over Ru supported on $ZnAl_2O_4$, CeO_2 and Al_2O_3, *Catalysis Today*, vol. 176, pp. 173-176, 2011.

[22] J. Okal, M. Zawadzki, W. Tylus, Microstructure characterization and propane oxidation over supported Ru nanoparticles synthesized by the microwave-polyol method, *Applied Catalysis B*, vol. 101, pp. 548-559, 2011.

[23] T. V. Choudhary, S. Banerjee, V. R. Choudhary, Catalysts for combustion of methane and lower alkanes, *Applied Catalysis A*, vol. 234, pp. 1-23, 2002.

[24] A. M. Gololobov, I. E. Bekk, G. O. Bragina, V. I. Zaikovskii, A. B. Ayupov, N. S. Telegina, V. I. Bukhtiyarov, A. Y. Stakheev, Platinum nanoparticle size effect on specific catalytic activity in *n*-alkane deep oxidation: Dependence on the chain length of the paraffin, *Kinetics and Catalysis*, vol. 50, pp. 830-836, 2009.

[25] H. Yoshida, Y. Yazawa, T. Hattori, Effects of support and additive on oxidation state and activity of Pt catalyst in propane combustion, *Catalysis Today*, vol. 87, pp. 19-28, 2003.

[26] S. M. Saqer, D. I. Kondarides, X. E. Verykios, Catalytic activity of supported platinum and metal oxide catalysts for toluene oxidation, *Topics in Catalysis*, vol. 52, pp. 517-527, 2009.

[27] T. Mitsui, K. Tsutsui, T. Matsui, R. Kikuchi, K. Eguchi, Catalytic abatement of acetaldehyde over oxide-supported precious metal catalysts, *Applied Catalysis B*, vol. 78, pp. 158-165, 2008.

[28] S. Hosokawa, Y. Fujinami, H. Kanai, Reactivity of Ru=O species in RuO_2/CeO_2 catalysts prepared by a wet reduction method, *Journal of Molecular Catalysis A*, vol. 240, pp. 49-54, 2005.

[29] S. Aouad, E. Saab, E. Abi-Aad, A. Aboukaïs, Reactivity of Ru-based catalysts in the oxidation of propene and carbon black, *Catalysis Today*, vol. 119, pp. 273-277, 2007.

[30] T. Mitsui, K. Tsutsui, T. Matsui, R. Kikuchi, K. Eguchi, Support effect on complete oxidation of volatile organic compounds over Ru catalysts, *Applied Catalysis B*, vol. 81, pp. 56-63, 2008.

[31] N. Kamiuchi, T. Mitsui, H. Muroyama, T. Matsui, R. Kikuchi, K. Eguchi, Catalytic combustion of ethyl acetate and nano-structural changes of ruthenium catalysts supported on tin oxide, *Applied Catalysis B*, vol. 97, pp. 120-126, 2010.

[32] T. Mitsui, T. Matsui, R. Kikuchi, K. Eguchi, Low-temperature complete oxidation of ethyl acetate over CeO_2-supported precious metal catalysts, *Topics in Catalysis*, vol. 52, pp. 464-469, 2009.

[33] S. Aouad, E. Abi-Aad, A. Aboukaïs, Simultaneous oxidation of carbon black and volatile organic compounds over Ru/CeO_2 catalysts, *Applied Catalysis B*, vol. 88, pp. 249-256, 2009.

[34] J. Okal, M. Zawadzki, Influence of catalyst pretreatments on propane oxidation over $Ru/\gamma-Al_2O_3$, *Catalysis Letters*, vol. 132, pp. 225-234, 2009.

[35] S. Bernal, J. J. Calvino, M. A. Cauqui, J. M. Gatica, C. Larease, J. A. Pérez Omil, J. M. Pintado, Some recent results on metal/support interaction effects in NM/CeO_2 (NM: noble metal) catalysts, *Catalysis Today*, vol. 50, pp. 175-206, 1999.

[36] G. A. Somorjai, F. Tao, J. Y. Park, The nanoscience revolution: merging of colloid science, catalysis and nanoelectronics, *Topics in Catalysis*, vol. 47, pp. 1-14, 2008.

[37] A. M. R. Galletti, C. Antonetti, S. Giaiacopi, O. Piccolo, A. M. Venzia, Innovative process for the synthesis of nanostructured ruthenium catalysts and their catalytic performance, *Topics in Catalysis*, vol. 52, pp. 1065-1969, 2009.

[38] J. Okal, Characterization and thermal stability of ruthenium nanoparticles supported on γ-alumina, *Catalysis Communications*, vol. 11, pp. 508-512, 2010.

[39] A. Miyazaki, K. Takeshida, K. Aika, Y. Nakano, Preparation of Ru nanoparticles supported on $\gamma-Al_2O_3$ and its novel catalytic activity for ammonia synthesis, *Journal of Catalysis*, vol. 204, pp. 364-371, 2001.

[40] M. Zawadzki, J. Okal, Synthesis and structure characterization of Ru nanoparticles stabilized by PVP or $\gamma-Al_2O_3$, *Materials Research Bulletin*, vol. 43, pp. 3111-3121, 2008.

[41] H. Y. H. Chan, Ch. G. Takoudis, M. J. Weaver, High-pressure oxidation of ruthenium as probed by surface-enhanced Raman and X-ray photoelectron spectroscopies, *Journal of Catalysis*, vol. 172, pp. 336-345, 1997.

[42] W. Zou, R. D. Gonzalez, Pretreatment chemistry in the preparation of silica-supported Pt, Ru, and Pt-Ru catalysts: An in situ UV diffuse reflectance study, *Journal of Catalysis*, vol. 133, pp. 202-219, 1992.

[43] J. Okal, The interaction of oxygen with high loaded $Ru/\gamma-Al_2O_3$ catalyst, *Materials Research Bulletin*, vol. 44, pp. 318-323, 2009.

[44] A. H. M. Batista, F. S. O. Ramos, T. P. Braga, C. L. Lima, F. F. de Sousa, E. B. D. Barros, J. M. Filho, A. S. de Oliveira, J. R. de Sousa, A. Valentini, A. C. Oliveira, Mesoporous MAl_2O_4 (M = Cu, Ni, Fe or Mg) spinels: Characterisation and application in the catalytic

dehydrogenation of ethylbenzene in the presence of CO_2, *Applied Catalysis A*, vol. 382, pp. 148-157, 2010.

[45] M. Zawadzki,W. Miśta, L. Kępiński, Metal-support effects of platinum supported on zinc aluminate, *Vaccum*, vol. 63, pp. 291-296, 2001.

[46] J. Okal, M. Zawadzki, Catalytic combustion of methane over ruthenium supported on zinc aluminate spinel, *Applied Catalysis A*, vol. 453, pp. 349-357, 2013.

[47] J. Okal, M. Zawadzki, K. Baranowska, Thermal treatment in air of the $Ru/ZnAl_2O_4$ catalysts for the methane combustion, *Applied Catalysis A*, vol. 471, pp. 98-105, 2014.

[48] S. Hosokawa, H. Kanai, K. Utani, Y. Taniguchi, Y. Saito, S. Imamura, State of Ru on CeO_2 and its catalytic activity in the wet oxidation of acetic acid, *Applied Catalysis B*, vol. 45, pp. 181-187, 2003.

In: Volatile Organic Compounds
Editor: Khaled Chetehouna

ISBN: 978-1-63117-862-7
© 2014 Nova Science Publishers, Inc.

Chapter 7

REMOVAL OF VOLATILE ORGANIC COMPOUNDS (VOCs) USING ADSORPTION PROCESS ONTO NATURAL CLAYS

H. Zaitan[*1] *and H. Valdés*[2]

[1]Laboratoire de Chimie de la Matière Condensée (LCMC),
Faculté des Sciences et Techniques,
Université Sidi Mohamed BenAbdellah, Fès, Maroc
[2]Laboratorio de Tecnologías Limpias (F. Ingeniería),
Universidad Católica de la Santísima Concepción, Concepción, Chile

ABSTRACT

Volatile organic compounds (VOCs) are critical toxic substances that may cause harmful effects on human health when emitted into the environment. The control of the emissions of VOCs into the atmosphere is presently one of the major environmental problems. Many conventional methods have been developed for industrial gaseous waste treatment but adsorption of contaminants onto adsorbents and their subsequent desorption for reuse or destruction has acquired high approval. Adsorption has been shown to be a cost-effective and environmentally friendly process compared to other technologies such as absorption, biofiltration, or thermal catalysis.

---

[*] Corresponding author: E-mail address: hicham.zaitan@usmba.ac.ma, hicham.zaitan@gmail.com (H. Zaitan).

This chapter summarizes the general backgrounds, described in the literature, related to the control of the emissions of VOCs. First, a state of the art of the sources of emissions of VOCs, and their main effects to human health and environment is presented. Then, the main technologies for VOCs control, their principles, limitations, applications and their remediation costs are briefly reviewed. Finally, the use of natural adsorbents as an economic alternative for the abatement of VOCs is discussed.

Keywords: Adsorption; oxidation; clays; Volatile Organic Compounds (VOCs)

1. INTRODUCTION

Air pollution, caused mainly by human activities, generates many levels of environmental impacts, such as: acid rain, ozone stratospheric depletion, increasing greenhouse effects or photochemical pollution. Since the Geneva Convention in 1979, many agreements and protocols have been signed in order to improve not only air quality, but also to prevent the causes of climate change. Over the last decades, enormous quantities of industrial pollutants have been released into the environment. Due to the high emissions of a wide variety of pollutants there has been a substantial increase on the number of environmental impacts and related health problems. The United Nations Environment Program estimates that about 3 million deaths occur each year worldwide due to an increase in air pollution levels.

Among the wide range of organic and inorganic air pollutants, volatile organic compounds (VOCs) deserve special attention, because of their impact on both human health and the global environment (Choung, 2001). VOCs are considered as the most common pollutants liberated by chemical industries involving mostly solvents, detergents, degreasers, thinners, cleaners, lubricants and liquid fuels. These chemicals are very harmful to both human health and the environment. Especially for human health, it is known that VOCs can often affect the central nervous system and may also have carcinogenic and mutagenic effects (Canet, 2007). Emissions of these compounds lead to the promotion of photochemical reactions in the atmosphere (ozone formation, stratospheric ozone layer depletion and formation of photochemical ozone smog) and odor nuisance (Calvert, 1994). With the increase of public concern about deteriorating air quality, more stringent regulations are being enforced to control air pollutants.

The reduction on concentration of these vapors from gas waste streams to acceptable levels is a serious challenge for the global chemical industry. Different physical-chemical technologies (recuperative and destructive processes), including adsorption, absorption, condensation, thermal and catalytic oxidation, and biological treatment processes are available. The choice of the best technology to control VOC emissions will depend on the actual operating conditions and the physical and chemical properties of organic compounds under use.

Adsorption techniques for the removal of VOCs is widely used in the field of organic pollution control with high efficiency, low cost, and convenient operation. In the process, the most important consideration is the selection of an appropriate adsorbent with sufficient efficiency to capture VOCs. Therefore, much attention has been focused on the use of new porous materials of high adsorption capacity, fast kinetics and high reversibility such as natural clays.

The aim of the present chapter is to summarize the state of the art of volatile organic compounds related to: (a) their source of emissions and their main effects on human health and the environment, (b) a description about the operating principles of physicochemical processes for VOC removal and (c) some examples of using adsorption onto natural clays for VOC control and their efficiency/cost performance.

2. DEFINITIONS AND CLASSIFICATIONS

Volatile organic compounds (VOCs) are a family of organic chemicals that are difficult to define. Until now, there is no standard or universal definition for a group classification of VOCs. Several classifications are provided taking into account both their physical-chemical properties such as pressure in combination with temperature, as their chemical reactivities (effect-oriented). According to the Solvents European Directive (1999/13/EC), VOCs are defined as organic compounds having a vapor pressure of 10 Pa or more at 20°C (VOC Solvents Directive, 1999). Another, effect-oriented definition used by the United States Environmental Protection Agency (US, EPA) in the Code of Federal Regulations 40 CFR 51.100(s) defines VOCs as any compound of carbon, excluding carbon monoxide, carbon dioxide, carbonic acid, metallic carbides or carbonates, and ammonium carbonate, which participates in atmospheric photochemical reactions (Moretti, 1993; U.S.EPA, 2009). The U.S. Environmental Protection Agency (EPA) has updated its definition of VOCs and now they are defined as organic chemical compounds whose composition makes them possible to evaporate under normal indoor atmospheric conditions of temperature and pressure (U.S.EPA, 2011). The U.S. National Research Council has described VOCs as organic compounds that vaporize easily at room temperature (USNRC, 2002).

However, in Europe there are other applicable definitions of VOCs to specific sectors. For example, the EU "Paint Directive" 2004/42/EC and European Eco-Labeling scheme (2002/739/EC amending 1999/10/EC) for paints and varnishes use boiling points and defines a VOC as an organic compound having an initial boiling point lower than or equal to 250°C at an atmospheric pressure of 101.3 kPa (Paints Directive, 2004; EPA, 2012). In Canada, the boiling points of VOCs were chosen roughly to be in the range of 50–250°C (Nathanson, 1995). According to the definition of the World Health Organization (WHO), VOCs are referred to as all organic compounds with a boiling point in the range of 50–260°C, excluding pesticides. In addition, the WHO categorizes organic pollutants as: very volatile (VVOCs), volatile (VOCs), and semi-volatile (SVOCs) (WHO, 1989; Crump, 2001; Ayoko, 2004). Table 1 includes three VOC Families usually under WHO classification.

Table 1. Classification of organic pollutants (adapted from WHO, 1989)

Description	Boiling Point Range (°C)	Examples
Very volatile (gaseous) organic compounds (VVOC)	<0 to 50-100	Propane, butane, methyl chloride
Volatile organic compounds (VOC)	50-100 to 240-260	Formaldehyde, d-Limonene, toluene, acetone, ethanol (ethyl alcohol) 2-propanol (isopropyl alcohol), hexanal
Semi-volatile organic compounds (SVOC)	240-260 to 380-400	Pesticides (DDT, chlordane, plasticisers (phthalates), fire retardants (PCBs, PBB))

Each definition covers a broad variety of organic compounds with diverse chemical properties; including paraffinic, olefinic and aromatic hydrocarbons; and various oxygen-, nitrogen-, sulphur- and halogen-containing compounds (Calvert, 1994, Hunter et al., 2000).

3. Sources and Impacts of VOCs

3.1. Nature and Sources of VOCs

There are mainly two sources categories of VOCs: natural or biogenic and anthropogenic (Kansal, 2009). Natural origins of VOCs include wetlands, forests, oceans and volcanoes (Kansal, 2009; Sahu, 2012). The estimated global VOCs biogenic emission rate is about 1150 Tg C/yr (see Table 2) (Guenther et al. 1995; Talapatra and Srivastava, 2011; Sahu, 2012). They represent 69% of the total emissions of VOCs while the rest (31%) is due to anthropogenic sources (Middelton et al., 1995).

Anthropogenic emission sources of VOCs into the atmosphere can be divided into two main sources: (a) mobile sources (transportation sector) and (b) stationary sources (production of solvent, industrial solvent use, industrial combustion processes and storage). Vehicle emissions are often the main source of VOCs in urban areas (Barletta et al., 2000). In the urban areas, emissions from anthropogenic sources are important compared to the rural areas where there is significant biogenic VOC production (Talapatra and Srivastava, 2011).

VOCs have been widely used as dissolving and cleaning agents in many industrial processes such as printing, film coating, manufacturing of magnetic tapes, electronic chips, among other applications. In general, VOCs include a wide range of chemical substances such as aromatics, various chlorinated hydrocarbons, perfluorocarbons, alcohols, organic acids, aldehydes, esters, ketones and aliphatics. It has also been recognized that VOCs are the main pollutants emitted by the chemical, petrochemical and related industries. The emission sources of VOCs can be classified into three groups: (1) use of organic solvents, (2) road transport, and (3) petrochemical and general chemical industry (Shim et al., 2006).

Approximately 235 millions tonnes of VOCs are released per year into the atmosphere from anthropogenic sources, mainly because thousands of them are produced and used in our daily lives (Guenther et al., 1995). For example, in the USA ca. 40% of the VOCs emissions are released from transportation activities and the remaining 60% result from stationary sources; being equally divided between fuel combustion, industrial manufacturing and solvent emissions (Horsley, 1993).

In the case of France, anthropogenic emissions are concentrated in large industrial or urban areas, which in the last years have become predominant. In 1996, 16–21% of emissions were from natural origin, against 84% of anthropogenic origin (Dueso et al. 1996). CITEPA registers VOC emissions in France since 1988 and publishes an annual report on air quality. France's global emissions of VOCs passed from 2661 kt in 1988 to 634 kt in 2012, decreasing by 76.2. Figure 1 shows the sources of VOC emissions (1988–2012), and Table 3 includes the amounts of released VOCs in France from different sectors (1988–2012) (CITEPA, 2013). As can be seen in Figure 2, the main emitting activities of VOCs in France are road transport (10.4%) and manufacturing (35.1%).

Table 2. Global estimation of VOCs (hydrocarbons) (Singh et al., 1992)

Source		Emission (Tg-C/Yr)
Anthropo-genic	(1) Transportation	22
	(2) Stationary sources	04
	(3) Industrial processing including natural gas production	17
	(4) Biomass burning, forest fires	45
	(5) Organic solvents	15
Total		103
Natural	**Oceanic**	
	(1) Light	5-10
	(2) C9–C28 hydrocarbons	1-26
	Terrestrial	
	(1) Microbial production	6
	(2) Emission from vegetation	1140
Total		1170
Total Emission		1273

Table 3. Evolution of pollutant emissions in France

Minimum registered	634 kt in 2012
Maximum registered	2661 kt in 1988
Emissions in 2012	634 kt
Evolution 2012/maximum	76.17%

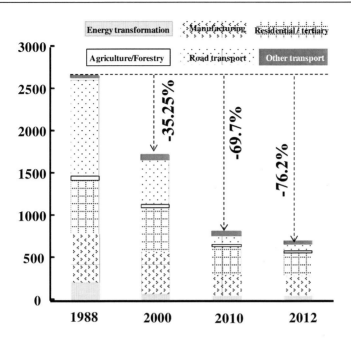

Figure 1. Emisions of VOCs by sector in France (kt).

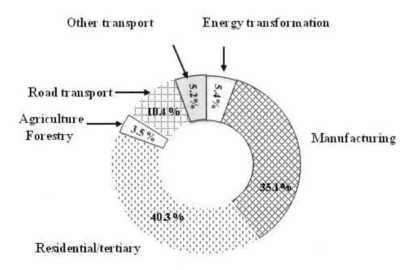

Figure 2. Emissions of VOCs by sector France (kt).

In 2012, the residential/tertiary sector (solvent use in domestic and building activities, biomass burning) and manufacturing (use of paints and solvents) were the first areas affected by VOC emissions, representing 75.4% of France's emissions. VOC emission distribution has changed since 1988 on the main three emitting sectors. On one hand, emissions generated by road transport have decreased by more than 90% (catalytic converters are more efficient and there is an increase on diesel vehicles) from the first to the third position. On the other hand, the industrial sector strongly produces the largest emissions of VOCs due to the use of solvents (degreasers, paints, printing and adhesive agents) together with the chemical and petrochemical industry. Agriculture and forestry are sectors that strongly emit VOCs as compared to the residential and tertiary economic sector. However, due to the large number of VOC sources and the extended nature of this economic activity, the control of these emissions are complicated. There has not been a decreasing record from this sector for ten years. The main VOCs emitted by the industry are divided into several chemical families. The VOC families that represent the most important emissions are alkanes and others such as methane ($\approx$26.50%), alcohols ($\approx$23.38%) and aromatics ($\approx$14.22%).

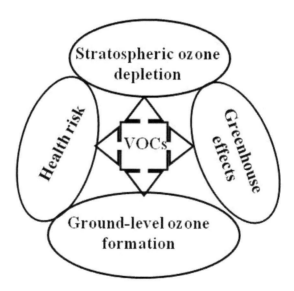

Figure 3. Effects of VOCs.

3.2. Effects of VOCs

VOCs are often used as solvents due to their ability to easily evaporate after use. These compounds have a lifetime in the range between 0.5 and 60 days in the atmosphere (Elichegaray, 2006). Pollution caused by VOCs do not only have immediate and local impacts (direct effects on humans) but also permanent and regional impacts, often with global repercussions (indirect effects on the environment). VOCs play a vital role on health and related environmental issues. They can cause a human health risk with long exposure time even at low concentrations; a global scale increase on VOCs enhances greenhouse effects; they cause stratospheric ozone depletion; and they are precursors of ground-level ozone formation (see Figure 3) (Demeestere et al., 2007; Sahu, 2012).

3.2.1. Health Effects

Pollution due to emissions of VOCs can be approached holistically following their direct effects (toxicological risks) or indirect (photochemical pollution) on man and on the receiving environment. These compounds have proved to cause a variety of adverse health effects (U.S.EPA; 2003). Their effects on human health vary from a simple nuisance to a serious hazard, leading to death (Hunter, 2000). Generally, nausea, emesis, epistaxis, fatigue, allergic skin reactions, dyspnea, decline in serum cholinesterase levels, eye and respiratory tract irritation, headaches, dizziness, and visual disorders are some of the symptoms associated with short-term exposure. Memory impairment is possible. Exposure may lead to leukemia. Damage to the nervous, reproductive and immune systems are caused by long-term exposure. Moreover, some VOCs, such as benzene, are known to have carcinogenic, teratogenic, and mutagenic effects (Canet et al., 2007). Finally, it has been indicated that short term exposure to ozone is associated with premature mortality following episodes of photochemical smog in the US (Volkamer et al., 2010).

3.2.2. Environmental Impacts

Among the effects of VOCs to the environment, two major concerns are related to air pollution problems: (a) stratospheric ozone depletion and tropospheric ozone formation.

3.2.2.1. Stratospheric Ozone Depletion

Stratospheric ozone (10-50 km above earth surface) forms a layer that protects the earth from the sun's ultraviolet radiation; protecting not only humans and animals from the risk of cancer and genetic mutations, but also the photosynthetic activity of plants. Many halogenated VOCs (e.g., tetrachloromethane, 1,1,1-trichloroethane, chlorofluorocarbons such as, CFC-11, CFC-12 and CFC-113) when they diffuse and break down into the stratosphere release chlorine atoms; disrupting the natural balance regulated by the presence of the ozone layer in the atmosphere due to UV photolysis reactions. They undergo numerous reactive processes ranging from: (i) hydroxyl radical reactions that lead to the release of active ozone-destroying chain carriers, and (ii) stratospheric photolysis (Molina et al., 1974). VOCs that contribute to ozone depletion are named ozone depleting substances. Those include bromine-containing fire retardants, fire extinguishers, chlorinated solvents and refrigerants.

3.2.2.2. Tropospheric Ozone Formation

Ozone (O_3) is naturally present in the atmosphere at a ratio of a few parts per million (ppm). Concentrations in the atmosphere are not constant. Indeed, the amount of ozone is higher in the stratosphere (10-50 km) than in the troposphere (0 to 10 km). Ground-level ozone is formed by the combination of molecule oxygen (O_2) with atomic oxygen (O) from the dissociation of nitrogen dioxide NO_2 as a result of sunlight. The destruction of ozone is due to its reaction with the nitrogen monoxide to nitrogen dioxide reforming. This cycle is called the Chapman cycle (see Figure 4).

Ozone is formed by a combination of atomic oxygen (O) with molecular oxygen (O_2), as follows:

$$O_2 + M + O \rightarrow O_3 + M \tag{1}$$

where M represents any other molecule which absorbs the energy of the reaction. Free unstable oxygen atoms result from the photolysis of nitrogen dioxide by sun light:

$$h.v + NO_2 \rightarrow NO + O \tag{2}$$

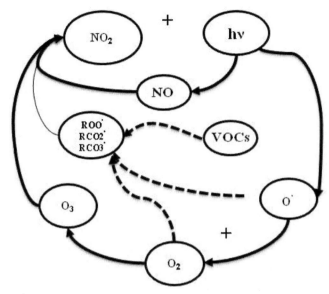

→Chapman cycle
--Peturbation of Chapman cycle (in the presence of VOCs)

Figure 4. Chapman cycle.

Another important reaction is that in which ozone reacts with nitric oxide. Ozone cannot be accumulated in the air because it will be consumed in the reaction represented by Eq. (3):

$$NO + O_3 \rightarrow NO_2 + O_2 \qquad (3)$$

In the presence of VOCs, the Chapman cycle is disturbed (see Figure 4). The role of VOCs can be summarized in the reduction on the amount of nitrogen monoxide, preventing reaction (3) to occur, as shown in Eq. (4).

$$VOC\text{–}OO + NO \rightarrow NO_2 + VOC\text{–}O \qquad (4)$$

Radicals generated by VOCs will react with NO to produce NO_2, which is the reactant for ozone production as mentioned in Eq. (1) and (2), and ozone will be growing up in the troposphere. Oxidized organic compounds are then mixed with different compounds and small particles in the air, creating photochemical smog, which has adverse effects on human health, vegetation and materials.

4. CONVENTIONAL TECHNIQUES FOR VOCs CONTROL

Various methods can be used to abate VOCs from gaseous emissions. The first step in the selection of a treatment method is to prepare an emission inventory. This inventory includes information and details on the entire industrial facility such as emitted pollutants, registered VOCs in each vent stream, periodical emission averages (including worst case emission scenarios), existence and status of pollution control equipment as well as regulatory status (Khan et al., 2000). The appropriate control device may be selected after consideration of a number of factors, for example, a regulation mandating specific control equipment for a particular VOC emission problem (Muzenda, 2012). However, most regulations set a maximum level of emissions without enforcing the abatement technology to be used. The use of an emission inventory allows selection of the appropriate abatement technique on the basis of various factors such as: cost, VOC inlet concentration, effluent gas flow rate and the required control level (Muzenda, 2012).

Available techniques for the abatement of VOCs can be divided in (a) process optimization by implementation of "green technologies" and (b) cleaning of flue gases (Rasu and Dumitriu, 2003). The reduction of VOC emissions can be made at different levels (act at the source or after the formation of pollutants) (US.EPA, 1999; Khan et al., 2000):

- applying a preventive or clean technology approach (Martin et al., 1992). This preventive approach eliminates emissions at the source by changing the processes or by technology modification, including substitution of raw materials to reduce VOC input to the process; changes in operating conditions in order to minimize solvent volatilization (using methods such as high performance electrostatic spraying), and the modification of equipment to reduce the escape of VOCs into the environment. However, changes in the current processes or products remain as a long-term procedure that can sometimes be very expensive (investments, changes). This approach often involves radical changes in processes and even adaptations can be relatively large (Dueso, 1997). When all preventive treatments are inadequate or impossible to implement, remedial actions are needed;
- applying a remedial approach: available technologies for the control of VOC emissions are divided into two types of techniques: recovery techniques and destructive techniques. Both could be complementary.

- by capture and recovery: such as changing operation units so that VOCs can be recovered and re-used in the process (this is possible using techniques like condensation, adsorption, absorption and membrane technology);
- by destruction: VOCs can be destroyed by combustion (incineration), catalytic oxidation or by microorganisms (biodegradation). In these cases, VOCs are transformed into CO_2 and H_2O.

Figure 5 shows a diagram of the major VOC control techniques currently in use.

The recovery of VOCs uses four main technologies, namely: absorption, adsorption, condensation and separation membranes.

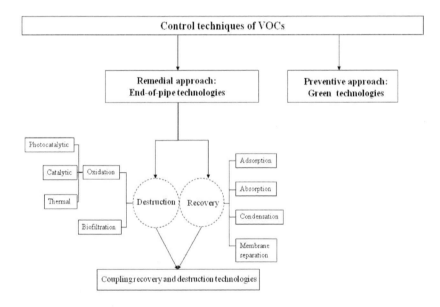

Figure 5. Classifications of VOC control techniques.

A non-destructive technique is absorption (scrubbing). This technique is used to separate and recover streams containing high concentrations of organics, especially water-soluble compounds. Absorption is a process consisting of the dissolution of a pollutant in a liquid (Le Cloirec, 1998; Khan et al., 2000; Parmar et al., 2008). In absorbers (or scrubbers), the vapor stream is introduced into an absorption chamber where it is mixed with the liquid.

VOCs are transferred from a gas stream to a liquid absorbent. The liquid must be treated to recover the pollutant and reuse the liquid or to dispose of the spent solvent if the absorbent cannot be regenerated. Absorption has several limiting factors, namely: the choice of a suitable solvent, and the VOC must be soluble in the absorbing liquid. Moreover, this technique requires regenerations of the solvent containing VOC by stripping. The VOC is then recovered in a condenser. The absorption processes can be dimensioned to allow a wide flow range (1000 to 100000 $Nm^3 \cdot h^{-1}$) and concentration (2 to 50 $g \cdot Nm^{-3}$). However, it can be difficult to treat effluents of low concentration and reuse the recovered products, which makes the economic reasons of little application for this technique in the treatment of VOCs.

As a matter of fact, condensation is a good technique for treating low-flow concentrated VOCs streams (Athken, 1995; Khan et al., 2000). Condensation can be achieved using a refrigeration system. The main problem is when highly volatile compounds are to be recovered; the cooling temperature should be very low. Among the various techniques of condensation, there are the mechanical condensation processors through exchangers and compressors (limited at temperatures from -30 to -40°C) and condensation at cryogenic liquid nitrogen (up to -180°C). The last technique is widely used in the pharmaceutical industry where nitrogen is used for inerting the reactor. This method is particularly suited to volumetric flow lower than 2000 $Nm^3 \cdot h^{-1}$ and at high concentrations (greater than 15 $g \cdot Nm^{-3}$). VOC concentrations in gaseous waste streams are generally too low for this technique to be effective.

Some VOCs can also be recovered from waste streams by membrane separation. In this process, the organic compounds present in the waste stream permeate the membrane, being separated from the treated air (Khan, 2000; Parmar et al., 2008). This technique should only be considered for low rate streams, when condensation or adsorption cannot achieve the desired levels of recovery efficiency.

Among the recovery techniques, adsorption is a good solution for treating industrial streams of low concentration of VOCs. Therefore, adsorption has been widely recognized as an effective means to control various emissions and, in some applications, for recovering reusable materials from exhaust streams (Hirota, 2004). Solvent recovery is a particularly common application of adsorption for controlling the emission of VOCs. An adsorption process allows treating effluent flow, which can range from 150 to 100 000 $m^3 \cdot h^{-1}$ for a range of concentration between 20 and 5000 ppmv. Process temperature inversely affects the adsorption capacity, and it usually takes place at room temperature (Le Cloirec, 2008).

In the adsorption process, the pollutant is collected on the surface (primarily the internal surface) of a granule or a crystal sorbent medium (Rhthven, 1984; Wang et al., 2001). It works conventionally in a fixed bed contactor and alternating operates two adsorbent beds with a regeneration system. The adsorbed compound is mainly physical retained, somewhat loosely, and can be released (desorbed) relatively easily by either heat or vacuum processes. Adsorption processes are used in 15% of cases of treatment of gaseous effluents containing VOCs. Adsorbers work at low temperatures, in order to favor the adsorption process.

Each adsorbent material has a different adsorption capacity referred to as the "adsorption isotherm." The adsorption capacity is measured in pounds of pollutant adsorbed per pound of adsorbent at a given temperature. Adsorption isotherm is a function of contaminant concentration (or partial pressure) in the vapor phase, temperature, ambient pressure, and the adsorbent surface area that VOCs can reach. Since activated carbons, zeolites, and polymers have different pore sizes and surface areas; the adsorption isotherm is different for each material and type of pollutant. These factors determine the amount of contaminant that each sorbent can adsorb. Selection of an appropriate adsorbent material is a function primarily of the contaminant to be adsorbed. However, the adsorption capacity of certain sorbents may be reduced by the relative humidity of the gas stream (Valdés et al., 2014).

Until now, activated carbons are the most effective adsorbent materials and are widely used in the adsorption process because of their large surface area, an extensive microporous structure, and a hydrophobic nature (Adolphs et al., 1996; Manjare et al., 2006). Many studies related to adsorption onto activated carbons have been performed due to the benefits mentioned above (Hwang et al., 1997; Shork et al., 2000; Yun et al., 2000; Bagreev et al., 2001; Huang et al., 2003; Ahn et al., 2004; Otero et al., 2005).

However, adsorption techniques only transfer the pollutants from one phase to another one, generating new waste with a higher content of VOCs. Due to its non-destructive character, the adsorbent must be regenerated after each use. The regeneration processes appear as a critical step to reduce costs and increase performance in the use of adsorbents. Regarding this problem, a variety of regeneration techniques have been evaluated and implemented.

Regeneration is performed by changing the conditions in the bed in order to desorb the VOCs from the adsorbent (Yang, 1987; Le cloirec, 1998; Bonjour et al., 2002; Lee et al., 2007; Lee et al., 2008; Yu et al., 2007). This process is achieved either by increasing the temperature using hot air or steam, or by decreasing the partial pressure, or introducing a stronger adsorbed material to displace the VOCs (Rafson, 1998). However, the regeneration of adsorbents such as activated carbons is very difficult. Fire issues and problems on the regeneration of high boiling point organic compounds have been reported (Blocki, 1993; Zhao et al., 1998; Baek et al., 2004; Makowski et al., 2007; Zaitan et al., 2008). Torrents et al. (1997) and Sheintuch and Matatov-Meytal (1999) report that the activated carbon pore structure is destructed and the smaller pores are clogged after few regeneration cycles.

In the last decades, hydrophobic synthetic zeolites and polymers are two other types of adsorbents which have been subjects of several studies (Yun et al. 1998; Jee et al. 2004; Cosseron et al. 2013). They have high adsorption capacities and a non-flammable character. They are not affected by high humidity levels. However, their production costs are significantly higher than activated carbons. Highly polar volatile compounds and degradation by-products of VOCs, such as vinyl chloride, formaldehyde, sulfur compounds, and alcohols, are better adsorbed by hydrophilic zeolites than by activated carbons. Polymeric adsorption is applicable to a wide range of VOCs and chlorinated VOCs at a wide range of vapor flow rates (P. Liu et al. 2009; C. Long et. al. 2012)

In recent years, research interest has been focused on the production of alternative adsorbents to replace costly activated carbons and zeolites. Attention has been directed to various natural microporous materials, which are able to remove pollutants from gaseous waste emissions at low cost. Cost is actually an important parameter for comparing adsorbent materials. An adsorbent can be considered as low-cost if it requires little processing and if it is abundant in nature. It may also be defined as by-products or waste materials from industry. Certain waste products from industries and agricultural operations, natural materials and bioadsorbents represent potentially economical alternative adsorbents. Many of them have been tested and proposed for VOCs removal such as bentonite and diatomite, dolomite, and natural zeolites, among others (Zaitan et al., 2005; Houari et al., 2007; Qu et al., 2009; Zuo et al., 2012; Alejandro et al., 2012; Valdés et al., 2014).

Different Adsorber Flow Arrangements

The main types of adsorber flow arrangements are fixed, moving, and fluidized beds. The fixed-bed is the most common adsorption process used in gas treatment. In fixed-bed systems, the gas flow is often conditioned by cooling upstream, partial condensation and heating to reduce the relative humidity. This is done to minimize co-adsorption of water. The adsorbent is contained within a cylindrical reactor, and the contaminated vapor is directed vertically downward or horizontally through the reactor. Normally there are two adsorbers working in parallel to allow continuous regeneration. One bed is adsorbing while the other is regenerating.

A fluidized-bed arrangement uses adsorbent beads that, by utilizing the velocity of the gas, are in a fluidized state throughout the tank. This is done with a continuous adsorption/desorption of the beads, removing the saturated beads from the bottom of the tank for regeneration and then adding them back to the system at the top. This technique is not very common due to the high energy consumption.

A continuous moving-bed process is much like the fluidized bed, except that the adsorbent does not float on the gas but is continuously being fed on top of the tank, passing the gas stream in counter-current. The saturated adsorbent is transferred to a moving bed regenerator at the bottom of the tank and once regenerated it is fed back at the top. In moving-bed systems, the adsorbent is continuously moving between two coaxial rotating cylinders, and the vapor flows between the two cylinders. As the cylinders rotate, part of the adsorbent is regenerated, while the rest continues to remove contaminants from the vapor stream.

When non-destructive techniques cannot be applied, waste streams containing VOCs should be treated using destructive processes. Thermal oxidation (incineration) represents 75% of currently installed facilities for VOC treatment. In this process, oxygen not only oxidizes VOCs to carbon dioxide and water by the increase in temperature, but other elements such as NOx, HCl, SO_2 are also oxidized. General chemical reaction for thermal oxidation of VOCs is as follows:

$$VOC + O_2 + heat \rightarrow CO_2 + H_2O + heat \tag{5}$$

Heat is required to achieve the necessary temperature for this reaction to occur. The required temperature is a function of several factors, including the presence of a catalyst. The heat shown on the right side of the equation represents the heat released by the reaction. Operating temperatures range from 650 to 1100°C. However, incomplete combustion is a permanent problem, leading to the formation of undesirable by-products (that are even more harmful than the input gas) such as furans and dioxins. Moreover, when the waste stream contains chlorinated VOCs, special care should be taken, since highly corrosive acid gases are formed.

There are two kinds of incinerators, being thermal or catalytic. They are distinguished by the energy recovery mode: recuperative incinerators (the waste gas is pre-heated with hot clean gas, leaving the combustion chamber) and regenerative incinerators (the heat is stored in a ceramic packed bed which will be used to pre-heat the incoming gaseous waste stream) (Techneron et al. 1987; Frost et al. 1991; Hermia and Vigneron, 1993; Van der Vaar et al. 1994; Thalhammer, 1995). These processes are suitable to reduce emissions from almost all sources of VOCs, with high efficiencies. However, heat recovery systems are only suitable for streams with high concentration of VOCs.

Another oxidative process for VOCs is catalytic oxidation. The catalyst increases the reaction rate, allowing the reaction to occur at lower temperatures (between 320 and 430°C) compared to thermal oxidation. Catalysis is a versatile and efficient technique not only for destroying VOCs but also for the elimination of other air pollutants, being known in this context as environmental catalysis. Catalytic oxidation offers high destructive efficiency, lower operating temperature than thermal combustion systems and smaller units. The main disadvantage of this process is that the catalyst is expensive. It has to be changed after a few years of operation and can be sensitive to poisoning by non-VOC materials present in the waste stream, such as sulphur. In addition, particulate matters in the waste gas, which can block catalyst pore structure, can affect the catalyst efficiency.

It is also possible to use biological treatment to destroy biodegradable volatile compounds. VOCs can be used as a carbon source. Microorganisms (bacteria, yeasts, fungi) transform organic pollutants to carbon dioxide and water. The process uses either a bioscrubber or a biofilter. This kind of process requires an adequate moisture level to work efficiently. Moreover, this technique is particularly suitable to treat odorous streams, to treat streams containing low molecular weight compounds and compounds like aldehydes, ketones, alcohols and organic acids. When waste streams contain halogenated and polyaromatic compounds, the biofiltration process is more difficult and not as efficient. Biological processes accept load fluctuations, and removal efficiency of higher than 90% can be obtained. Biofiltration offers several advantages over other control technologies. Since natural media rather than synthetic materials are used in the filter beds; capital cost is strongly influenced by the size of the biofilter. Capital and operating costs for biofiltration are also minimized by the fact that oxidation of VOCs is achieved by microorganisms, rather than by chemicals or heat. It does not create issues of potentially hazardous waste to be disposed of, so it is known as a "green" technology. However, the use of biofiltration to remove VOCs still has some limitations and challenges. Biofilter limitation applications include low treatment efficiency at high concentrations of pollutants, extremely large size bioreactors, space constraints require close control of operating conditions, and limited life time of packing bed due to clogging of the medium by a high concentration of biosolids (Delhomenie and Heitz, 2005, Kumar et al., 2011).

4.1. Cost and Performance

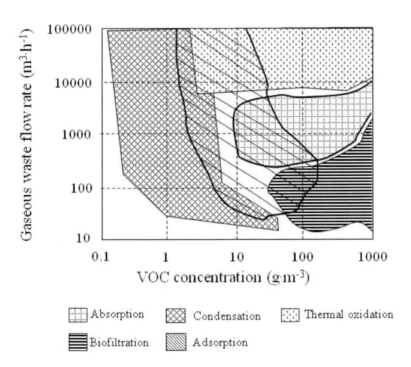

Figure 6. Economic feasibility zones of various processes for treating VOCs.

In spite of the fact there are numerous technologies for the emission control of VOCs, all are not applicable everywhere. Often, the two main selling points of a method of treatment is its effectiveness and cost (both investment and operation costs). The most used technologies are shown in figure 6. Figure 6 depicts economic feasibility zones of various processes for treating VOCs, depending on the concentration and flow rate to be treated.

Table 4 compares various available technologies for VOC control. All technologies have their own applicability depending upon the source, type and concentration of VOC (Bohn, 1992; Ademe, 1997; Dueso, 1997; Khan et al., 2000; Parmar et al., 2008).

Table 4. Current technologies for air pollution control

Methods (Conventional and upcoming)	Technology	Operational characteristics			Advantages	Limitations
		Gas flow $(Nm^3.h^{-1})$	Temperature $(^{\circ}C)$	VOC $(g.Nm^{-3})$		
Adsorption	Activated carbons, zeolites, alumina, clays,..	10^3-100×10^3	<55	1 -50	Proven and efficient, Low capital investment, Good for solvent recovery	Selective applicability Moisture and temperature constraints Adsorbant is too specific and can saturate fast; Risk of pollutant reemission
Absorption	Washing gas with contaminated water	10^3-100×10^3	Normal	1-30	Especially good for inorganic acid gasses, Possible recovery of VOC	Not suitable for low Concent-rations, generates wastewater
Condensation	Liquefaction by cooling or Compression	$<2 \times 10^3$	Ambient	>10	Good for product or solvent recovery	Further treatment is required, Applicable in high concentrations only
Recovery Techniques — Filtration	Air passed through fibrous material coated with viscous materials	100-10×10^3	10-41	>60	Efficient for particle removal, Compact and commonly used	Unable to remove gases, fouling, particle reemission can occur due to microbial growth.
Membrane Separation	Separation through semi permeable membranes	5-100	Ambient	>50	Recommended for highly loaded Streams	Membrane fouling and high pressure is needed
Destructives Techniques — Incineration	Thermal oxidation	1×10^3-300×10^3	750 -800	1- 12	High destruction efficiency	Not cost effective, incomplete mineralization and release of secondary pollutants.

Removal of Volatile Organic Compounds (VOCs) ...   189

Methods (Conventional and upcoming)	Technology	Operational characteristics			Advantages	Limitations
		Gas flow (Nm^3.h^{-1})	Temperature (°C)	VOC (g.Nm^{-3})		
Catalytic Oxidation	Thermal catalysts (Pt, Al, ceramics)	1×10^3- 100×10^3	250-400	1-12	Efficient, Conserves energy	Catalyst deactivation and its disposal, formation of byproduct
Photolysis	UV radiations to oxidize air pollutants and kill pathogens	--	Normal	--	Removes fumes and gaseous pollutants	Release of toxic photoproducts, UV exposure may be hazardous and energy consuming.
Photo catalysis	High energy UV radiation used along with a photocatalyst	--	--	--	Energy intensive, Popular method suitable for broad range of organic pollutants	Exposure to UV radiation may be harmful
Ozonation	Strong oxidizing agent	--	--	--	Removes fumes and gaseous pollutants	Generates unhealthy ozone and degradation products.
Microbial abatement	Air passed through a packed bed colonized by attached microbes as biotrickling filters or microbial cultures in bioscrubbers	$<150\times10^3$	--	<5	Cost effective, more efficient, eco-friendly,	Need for control of biological parameters

(Destructives Techniques)

As seen in Table 5, investment and operating costs fluctuate according to technical performances. Overall, the most expensive techniques both in investment and operation are oxidation processes, ensuring good performance (Manero and Roustan, 1995; Le Cloirec, 1998; Khan et Ghoshal, 2000; Le Cloirec et al., 2003; USEPA, 2006). Biofiltration is the more financially attractive technique. However, it only applies to specific cases (in the presence of soluble VOCs).

Table 5. Average performance and costs of treatment techniques

Device	Efficiency (%)	Investment cost (€/m³.h)	Operating costs (€/1000 m³)
Catalytic oxidation	90 - 98	5 - 200	1 - 10.33
Thermal oxidation	95 - 99	5 - 96	0.6 - 6.2
Adsorption	80 - 90	7 - 32	1.7 - 8.2
Absorption	95 - 98	7 - 55	0.7 - 2.4
Condensation	50 - 90	5 - 37	1.4 - 8.2
Biofiltration	> 90	15 - 25	0.2 -0.5

Other emerging technologies (advanced oxidation processes (AOPs) are subjects of increased research and are currently being evaluated, such as ultraviolet (UV) oxidation technology (Guo et al. 2008; Sleiman et al., 2009), catalytic ozonation (Kwong et al. 2008, Brudo, 2011; Einaga et al., 2011) photolysis (Chang, 2009), photocatalysis (Monneyron, 2004) and plasma technology (Huang et al., 2011.). UV oxidation uses oxygen-based oxidants like ozone or peroxide to convert VOCs into carbon dioxide and water in the presence of UV light. This technique can also be used together with an adsorbent, like activated carbons, which removes the unreacted VOCs from the process. The other technology uses plasma (a high temperature ionized gas that is very reactive), which can initiate dissociation reaction in VOC molecules. Decomposition reactions are initiated by free-radical mechanism. This technique is highly selective for the decomposition of halogenated hydrocarbons. Moreover, the reaction temperature is much lower than the ones used in catalytic oxidation (Naydenov and Mehandjiev 1993; Xi et al., 2005).

4.2. Selection Criteria

The first step is to determine the nature of VOCs to be treated, their minimum and their maximum average concentrations. It is also important to register additional parameters of the gaseous waste stream under evaluation such as: temperature, relative humidity and the presence of other air pollutants.

The choice of a treatment method (or maybe a coupling of different techniques) will depend on the following criteria: desired level of removal efficiency, physical-chemical characteristics of VOCs (e.g., stickiness, volatility, molecular weight, among others), properties of gaseous waste stream (concentration, flow rate, heat content, vapor pressure, temperature, particulate matter and moisture content), safety issues (flammability, explosivity), and the value of recovered material. Moreover, it is important to combine the sought after treatment objectives with opportunities for energy and raw material recovery; and to evaluate the impact of the selected technology on upstream processes and on the generation of secondary contaminants.

Table 4 summarizes performances, advantages and disadvantages of different technologies for VOC treatment. However, there is not a unique and obvious method to select for reducing VOC emissions. It is therefore difficult to say today which technologies will prevail for VOC control in the future.

4.3. Non-Conventional Low-Cost Adsorbents for the Removal of VOCs

Natural clay minerals are familiar to mankind from the earliest days of our civilization. Natural clays might be considered for the abatement of VOCs from gaseous waste streams because of their low cost, abundance in most continents of the world, high adsorption properties and potential for ion-exchange. Their high adsorption capacities are related to their high-surface-area and high porosity. There have been some published reports about the adsorption of VOCs on natural clays (Zaitan et al., 2005; Houari et al., 2007; Qu et al., 2009; Zuo et al., 2012). Bentonite clay from the Nador province in Morocco has been registered as the one with the largest surface area $(83.5 \text{ m}^2\text{g}^{-1})$. Its current market price (about US$ 0.04$ kg^{-1}) is considered to be 250 times cheaper than activated carbons (Zaitan et al. 2008). In recent years, there has been an increased interest in applying clay minerals such as bentonite and diatomite earth as adsorbents for the removal of VOCs. In this section, adsorptive capacities of natural clays (Bentonite (BT) and diatomite (DT)) toward VOCs are compared, taking into account the cost/efficiency ratio and the ability of natural clays to be easily regenerated.

4.3.1. Natural Clays Compared to Commercial Oxides

Adsorption isotherms of xylene on different adsorbents in a concentration range between 0 - 104 ppmv and at a temperature between 300 and 373 K are shown in Fig. 7. Adsorption equilibrium data are fitted using Langmuir and Freundlich models.

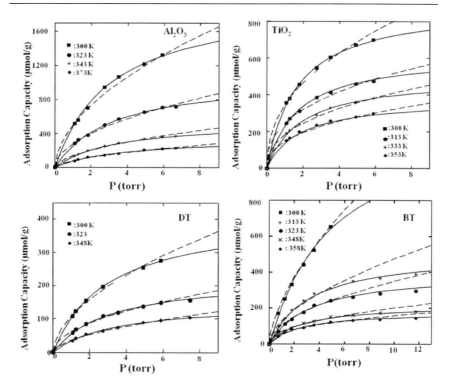

Figure 7. Experimental data (symbols). Langmuir isotherms (solid lines) and Freundlich isotherms (dashed lines) of xylene adsorption on Al_2O_3, TiO_2, DT and BT.

Maximum adsorption capacities, Q_m (µmole·g^{-1}), at 300 K lead to the following decreasing order: SiO_2 >Al_2O_3> BT>TiO_2>DT. This is clearly related to specific surface area (S_{BET}) and total porous volume, including mesopore and micropore. A calculated number of molecules of xylene per gram and per surface of adsorbent that form a monomolecular layer are given in Fig. 8. Additionally, a summary of maximum adsorption capacities of various natural and commercial adsorbents toward xylene is listed in Table 6.

BET surface areas of the natural clays and commercial oxides are 83.5, 21, 107, 55 and 200 m^2.g^{-1} for BT, DT, Al_2O_3, TiO_2, and SiO_2, respectively. All samples exhibit an overall mesoporous structure with an average pore diameter ranging from 6 to 11 nm. Experimental results of xylene adsorption onto bentonite show a maximum adsorption capacity of 420 µmole. g^{-1} at 0.36 kPa partial pressure of xylene. This value is significantly higher than the obtained values using other adsorbents under similar experimental conditions such as silicates, MCM-22 zeolite (Corma et al., 1996), Webster soil (Pennell et al.,

1992), kaolinite (Pennell et al., 1992) and Webster HP (Pennell et al., 1992). Although other microporous materials have significantly higher adsorption capacity than DT, such as PCH (Qu et al., 2009), MIL-101 (Zhao et al., 2011), activated carbon AC40 (Benkhedda et al., 2000), CA (Wang et al., 2004), Al-Meso 100 (Huang et al., 2006), UL-ZSM5-100-2 (Huang et al., 2006), UL-ZSM5-100-6 (Huang et al., 2006), silicalite (Talu et al., 1989), ITQ-1 zeolite and AlPO4-5 (Chiang et al., 1991), which have shown adsorption capacities of 2000, 9600, 4666, 1000, 2800, 1800, 1800, 2050, 1170, 801 and 1166 μmole xylene·g^{-1}, respectively. These higher values are related to higher specific surfaces of these solids. However, the value of the adsorption capacity obtained for BT sample (in μmole·m^{-2}) is about 1-5 times greater than those of MIL-101 , AC40, CA, and synthetic zeolites as Al-Meso 100, UL-ZSM5-100-2, ULZSM5-100-6 as compared to some data given in the literature for xylene adsorption on different kinds of zeolites at 300 K (Chiang et al., 1991; Huang et al., 2009) and activated carbons (Benkhedda et al., 2000) . Reported values of adsorption capacities of xylene isomers on NaY, KY, BaY and AlPO4-5 zeolites, are 2164, 2011, 1801, and 832 μmole·g^{-1}, respectively. However, the values obtained with bentonite are higher than the values of other natural minerals such as Webster soil (2.6 $m^2·g^{-1}$), Webster HP (33 $m^2·g^{-1}$), Kaolinite (13.6 $m^2·g^{-1}$), or silica gel (238 $m^2·g^{-1}$) that give maximum adsorption capacities toward xylene of 53.6, 116, 42, and 609 μmole·g^{-1} (corresponding to 20.6; 3.5; 3.1; 2.55 μmole·m^{-2}), respectively

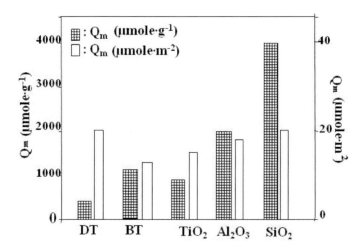

Figure 8. Maximum adsorption capacity (by Langmuir model).

Table 6. Maximum adsorption capacities of different adsorbents toward xylene

Adsorbent	VOCs	Temperature (°C)	Maximum adsorption capacities (µmol/g)	Reference
SiO2 Al2O3 TiO2 DT BT	xylene	27	4000 2000 890 420 1042	(Zaitan et al., 2013)
NaY	p-xylene	25	2164	(Simonot-Grange et al., 1997; Pilverdier, 1995; Bellat, 1995[a])
KY	p-xylene	25	2011	(Simonot-Grange et al., 1997; Pilverdier, 1995; Bellat, 1995)
BaY	p-xylene	25	1778	(Simonot-Grange et al., 1997; Pilverdier, 1995; Bellat, 1995[a])
BaX	p-xylene	50	1428	(Simonot-Grange et al., 1997; Pilverdier, 1995; Bellat, 1995)
ZSM-12	m-xylene	40	$\approx$600	(Guil et al., 1998)
AC40	m-xylene	25	6290	[35] Benkhedda et al., 2000)
AC1 AC2 AC3	o-xylene	30	1094 2061 3001	(Huang et al., 2009)
Actif Carbon (CA)	o-xylene	15 25	3600 3500	(Simonot-Grange et al., 1997)
MCM-22	o-xylene	42	320	Corma et al., 1996)
Al-Meso-100 UL-ZSM5-100-2 UL-ZSM5-100-6 UL-ZSM5-100-8 ZSM5-100	o-xylene	$30^{(T1)}/50^{(T2)}/80^{(T3)}$ $30^{(T1)}/50^{(T2)}/80^{(T3)}$ $30^{(T1)}/50^{(T2)}/80^{(T3)}$ $30^{(T1)}/50^{(T2)}/80^{(T3)}$ $80^{(T2)}/100^{(T3)}$	$12830^{(T1)}/9920^{(T2)}/6740^{(T3)}$ $5300^{(T1)}/4840^{(T2)}/4230^{(T3)}$ $5800^{(T1)}/4320^{(T2)}/2770^{(T3)}$ $840^{(T2)}/690^{(T3)}$ $910^{(T2)}/690^{(T3)}$	Huang et al., 2006
HZSM5-/180	p-xylene	27	2116	Meininghaus et al., 2000
NaZSM5-/180	p-xylene	27	2030	
StY2-L	p-xylene	27	3190	
SiCl4Y2-L	p-xylene	27	3830	
HMOR	p-xylene	27	2010	
Silicate-1 (B)	p-xylene	50 75	$\approx$800 $\approx$700	Song et al. 2000
ZSM-5	p-xylene	50	$\approx$1100	
DAY zeolite	m-xylene	45.5 55.5	$\approx$ 1330 $\approx$ 1320	Brihi et al., 2003
MOF 1 (metal–organic framework)	o-Xylene p-Xylene m-Xyelene	125	$\approx$320	Bárcia et al., 2012
MIL-101	p-xylene	25	9670	Zhao et al., 2011
MIL- 53(Al)	o-xylene	$30^{(T1)}/60^{(T2)}$	$4000^{(T1)}/3800^{(T2)}$	Duan et al., 2013
CoAlPO4-5	p-Xylene	69	624	Liu et al., 2004
sludge-based	o-xylene	25	2330	Fang et al., 2012

Removal of Volatile Organic Compounds (VOCs) ...

Adsorbent	VOCs	Temperature (°C)	Maximum adsorption capacities (μmol/g)	Reference
Clay (PCH)	o-xylene	25	1840	Qu e al., 2009

The higher BET surface area of activated carbons AC40, AC1, AC2, and AC3 (1330, 321.4, 501.7 and 839 $m^2 \cdot g^{-1}$), make them have the highest adsorption capacity of 6290, 1094, 2061 and 3001 $\mu mole \cdot g^{-1}$, corresponding to ~5, 3.4, 4.11 and 3.57 $\mu mole \cdot m^{-2}$ for AC40, AC1, AC2, and AC3, respectively (Benkhedda et al., 2000; Huang et al., 2009). Those values are about 2-4 times lower than the values obtained for BT ($\approx$12.5 $\mu mole \cdot m^{-2}$), for alumina ($\approx$18.7 $\mu mole \cdot m^{-2}$) for silica and DT ($\approx$20 $\mu mole \cdot m^{-2}$), and for TiO_2 ($\approx$16.2 $\mu mole \cdot m^{-2}$), being consistent with data reported in the literature for other porous adsorbents under similar experimental conditions. Table 5 shows that adsorption capacities are about 1–2 times different as compared to those values obtained using zeolites, active carbon, and silicates (silicalite-1B).

After xylene adsorption equilibrium was reached onto bentonite, isothermal desorption under nitrogen flow was studied. The amount of xylene evolved during the isothermal desorption at ambient temperature was 370 $\mu mole$ of xylene.g^{-1} of bentonite, corresponding to 88% of the amount at the adsorption equilibrium. Similar experiments have been carried using DT, Al_2O_3 and TiO_2, and desorption yields of xylene are approximately 60-85% of the amount adsorbed at equilibrium. All the assessed solids allow a significant isothermal desorption at 300 K of the adsorbed xylene; indicating that it is weakly adsorbed for the most part, leading to an advantage in terms of easier regeneration. These findings are confirmed by the adsorption heat of xylene adsorption on DT, BT, TiO_2 and Al_2O_3. Isosteric adsorption heats are 1–1.5 times the heats of vaporization for xylene (43.43 $kJ \cdot mole^{-1}$) (Bellat et al., 1995).

4.3.2. Analysis of Adsorbent Costs

Nowadays, the development of an efficient and low-cost adsorbent is one of the challenging targets of researchers, focusing on adsorption processes. Table 7 shows a comparison of adsorption capacities of natural and commercial materials, expressed either in $mole \cdot kg^{-1}$ or $mole \cdot m^{-3}$, and taking into account adsorbent prices. As it can be seen, adsorption capacity of silica and alumina at 300 K (expressed in $mole \cdot kg^{-1}$), are almost 4 and 2 times higher than the values of BT, while these values are 9.5 and 4 times higher than DT.

Table 7. Adsorption capacities and prices of natural and commercial materials

Adsorbent	S_{BET} (m^2/g)	Apparent density (kg·m^{-3})	Maximum adsorption capacity (mole·kg^{-1})	Maximum adsorption capacity (mole·m^{-3})	Price (US$·kg^{-1})
SiO$_2$	200	50	4	200	1-10
Al$_2$O$_3$	107	50	2	100	52
TiO$_2$	55	35	0.89	31.2	--
DT	21	326	0.42	137	0.25
BT	83.5	550	1.042	573	0.04

As for the adsorptive capacity of the alumina, the values (expressed in mole·kg^{-1} or mole·m^{-3}) are 2 and 4 times greater as compared to BT and DT, respectively. This means that in order to have an equivalent adsorption capacity it is necessary to use a mass of bentonite 4 times higher than that required for silica. Nevertheless, taking into account the apparent density of the solid (see table 7), adsorption capacities for xylene (in mole·m^{-3} of adsorbent bed), is 3 times better in BT as compared to silica, 6 times relative to the Al$_2$O$_3$, 4.2 times for DT and 18 times relative to TiO$_2$. This indicates that the need of the adsorber size is 3 times smaller using bentonite clay than that required with silica (similarly 4 times smaller than with alumina and 18 times smaller with TiO$_2$). Therefore, the use of bentonite might help greatly in reducing the size of the adsorbent bed for similar adsorption efficiency. This is considered to be a significant advantage of bentonite over the other adsorbents showing its potential interest for captured xylene and other VOCs present in industrial effluents. Moreover, considering that an estimated average price of bentonite is around US$0.04·kg^{-1}, yields an additional advantage in terms of cost which is 1300 times lower than that of alumina (US$ 52·kg^{-1} and 500 times lower than of the activated carbon (US$20–22·kg^{-1}) (Zaitan et al., 2008). Thus, natural clay is likely to become a strong adsorbent candidate for VOC removal.

CONCLUSION

This work covers a wide range of adsorbents normally used for the removal of VOCs from gaseous waste streams. Inexpensive, locally available and effective materials could be used instead of commercially activated carbons for the removal of VOCs. Undoubtedly low-cost adsorbents offer many promising benefits for commercial purposes in the future. In particular, from the literature reviewed, several adsorbents such as treated bentonite and diatomite have demonstrated good removal capabilities for certain VOCs in comparison to activated carbons. Regarding adsorption performance, bentonite clay provides a promising potential for reducing the size of the adsorption facility. In addition, it presents the possibility of conducting easier regeneration and solvent recovery systems. This is of very significant interest from the point of view of design and operation of an adsorption facility.

ACKNOWLEDGMENT

Authors gratefully acknowledge CNRST Morocco/CNRS France (Project SPI 05/13) and CONICYT, FONDECYT/Regular (Grant No. 1130560) for their financial support.

REFERENCES

A.D.E.M.E, La *Réduction des Emissions de Composés Organiques Volatils dans l'Industrie,* 1997.

Adolphs, J., Setzer, J.M., A model to describe adsorption isotherm, *Journal of Colloid and Interface Science*, vol. 180, pp. 70-76, 1996.

Ahn, H., Lee, C.H., Effects of capillary condensation on adsorption and thermal desorption dynamics of water in zeolite 13X and layered beds, *Chemical Engineering Science*, vol. 59, pp. 2727–2743; 2004.

Alejandro, S., Valdés, H., Manero, M.H., Zaror, C.A., BTX abatement using Chilean natural zeolite: The role of Bronsted acid sites, *Water Sci. Technol.* 66 (2012) 1759-1769.

Athken, N., VOC's recovery using cooling and cryogenic processes: state of the art, main issues, test rig. *Odors and VOCs Journal*, vol. 1, pp. 251, 1995.

Ayoko, G. A., Volatile organic compounds in indoor environments. In: Hutzinger O, editor. *The handbook of environmental chemistry.* Springer-Veriag, pp. 1–35, 2004.

Baek, S. W., Kim, J-R., Ihm, S.-K., Design of dual functional adsorbent/catalyst system for the control of VOC's by using metal-loaded hydrophobic Y-zeolites, *Catalysis Today*, vol. 93–95, pp. 575–581, 2004.

Bagreev, A., Rahman, H., Bandosz, T.J., Thermal regeneration of spent activated carbon previously used as hydrogen sulfide adsorbent, *Carbon*, vol. 39, pp. 1319–1326, 2001.

Bárcia, P. S., Nicolau, M.P.M., Gallegos, J.M., Chen, B., Rodrigues, A.E., Silva, J.A.C., Modeling adsorption equilibria of xylene isomers in a microporous metal–organic framework, *Microporous Mesopourous Materials*, vol.155, pp. 220–226, 2012.

Barletta, B., Meinardi, S., Rowland, F.S., Chan, C-Y., Wang, X., Zou, S., Chan, L. Y., Blake, D.R., Volatile organic compounds in 43 Chinese cities, *Atmospheric Environment*, vol 39, pp. 5979–5990, 2005.

Bellat, J. P., Simonot-Grange, M.H., Jullian, S., Adsorption of gaseous p-xylene and m-xylene on NaY, KY, and BaY zeolites: Part 1. Adsorption equilibria of pure xylenes, *Zeolites, vol.* 15, pp. 124–130, 1995[a].

Bellat, J. P., Simonot-Grange, M.H., Jullian, S., Adsorption of gaseous p-xylene and m-xylene on NaY, KY, and BaY zeolites: Part 2. Modeling enthalpies and entropies of adsorption, *Zeolites*, vol. 15 (3), pp. 219–227, 1995[b].

Benkhedda, J., Jaubert, J.N., Barth, D., Perrin, L., Bailly, M., Adsorption isotherms of m-xylene on activated carbon: measurements and correlation with different models, *Journal of Chemical Thermodynamics*, vol. 32 (3), pp. 401–411, 2000.

Blocki, S. W., Hydrophobic Zeolite Adsorbent: A proven advancement in solvent separation technology, *Environmental Progress*, vol. 12, pp. 226-230, 1993.

Bohn, H., Considering biofiltration for decontaminating gases, *Chemical Engineering Progress*, vol. 88, pp. 34–40, 1992.

Bonjour, J., Chalfen, J.B., Meunier, F., Temperature Swing Adsorption Process with Indirect Cooling and Heating. *Industrial Engineering Chemistry Research, vol.* 41, pp. 5802-5811, 2002.

Brodu, N., Zaitan, H., Manero, M-H., Pic, J.-S., Removal of volatile organic compounds by heterogeneous ozonation on microporous synthetic alumino silicate, *Water Science Technology*, vol. 66, pp.2020–2026, 2012

Calvert, J. G., *Chemistry for the 21st Century*. The Chemistry of the Atmosphere: Its Impact on Global Change, Blackwell Scientific Publications, Oxford, 1994.

Canet, X., Gilles, F., Su, B.-L., de Weireld, G., Frere, M., Adsorption of alkanes and aromatic compounds on various faujasites in the Henry domain. 1. Compensating cation effect on zeolites Y, *Journal Chemical Engineering Data, vol* 52, pp. 2117–2126, 2007.

Chang, C. C., Chiu, C.Y., Chang, C.Y., Chang, C.F., Chen, Y-H., Ji, D-R., Yu, Y-H., Chiang, P-C., Combined photolysis and catalytic ozonation of dimethyl phthalate in a high-gravity rotating packed bed. *Journal of Hazardous Materials*, vol. 161, pp. 287-293, 2009

Chiang, A.S.T., Lee, C.K., Chang, Z.H., Adsorption and diffusion of aromatics in AlPO4-5, *Zeolites*, vol. 11, pp. 380–386, 1991.

Choung, J. H., Lee, Y.X., Choi, D.K., Adsorption Equilibria of Toluene on Polymeric Adsorbents, *Journal of Chemical Engineering Data*, vol 46, pp. 954- 958, 2001.

Corma, A., Corell, C., Pérez-Pariente, J., Guil, J.M., Guil-Lopez, R., Nicolpoulos, S., Gonzalez Galbet, J., Vallet-Regi, M., Adsorption and catalytic properties of MCM- 22: the influence of zeolite structure, *Zeolites* vol. 16, pp. 7–14, 1996.

Cosseron, A.F., Daou, T.J., Tzanis, L., Nouali, H., Deroche, I., Coasne, B., Tchamber, V., Adsorption of volatile organic compounds in pure silica CHA, ÆBEA, MFI and STT-type zeolites, *Microporous and Mesoporous Materials,* vol. 173, pp. 147–154, 2013.

Crump, D., Strategies and protocols for indoor air monitoring of pollutants. *Indoor Built. Environ*, vol 10, pp. 125–131, 2001

Delhomenie, M.C., Heitz, M., Biofiltration of air: A review. *Critical Reviews in Biotechnology*, vol. 25, pp. 53-72 2005.

Demeestere, K., Dewulf, J., De witte, B., Langenhove, H.V., Sample preparation for the analysis of volatile organic compounds in air and water matrices, *Journal of Chromatography A*, vol. 1153, pp. 130-144, 2007.

Duan, L., Dong, X., Wu, Y., Li, H., Wang, L., Adsorption and diffusion properties of xylene isomers and ethylbenzene in metal–organic framework MIL-53(Al), *Journal Porous Materials,* vol. 20 (2), pp. 431–440, 2013.

Dueso, N., Naublanc, J., *La réduction des émissions de COV dans l'industrie* (ed. l'Environnement, M. d.), 1996.

Dueso, N., La Pollution par les COV: Définitions, Sources et Solutions, *Informations Chimie*, N°387, p 76-79, 1997.

Einaga, H., Teraoka, Y., Ogata, A., Benzene oxidation with ozone over manganese oxide supported on zeolite catalysts. *Catalysis Today,* vol. 164, pp. 571–574, 2011

El Brihi, T., Jaubert, J-N., Barth, D., Perrin, L., VOCs isotherms on day zeolite by static and dynamic methods: experiments and modelling, *Environmental Tehnology*, vol. 24 (10) pp. 1201–1210, 2003.

Elichegaray, C., Pollution atmosphérique. *Techniques de l'Ingénieur*, G1500 v2, 2006.

Fang, P., Cen, C., Zhang, H., Tang, Z., Chen, D., Chen, Z., Chen, Z., Removal of VOCs using sludge-based adsorbents, *Advanced Materials Research*, vol. 599, pp. 305–308, 2012.

Freundlich, H., Uber die adsorption in losungen, *Z. Physical Chemical*. Vol. A57, pp. 385, 1906

Frost, A.C., Sawyer, J.E., Summers, J.C., Shah, Y.T., Dassori, C., Kinetics and transport parameters for the fixed-bed catalytic incineration of volatile organic compounds, *Environmental Science and Technology*, vol. 25, pp. 2065-2070, 1991.

Guenther, A., Hewitt, C.N., Erickson, C.D., Fall, R., Geron, C., Graedel, T., Harley, P., Klinger, L., Lerdau, M., McKay, W.A., Pierce, T., Scholes, B., Steinbrecker, R., Tallamraju, R., Taylor, J., Zimmerman, R., A global model of natural volatile organic compound emissions, *Journal of Geophysical Research: Atmpsphers*, vol 100, pp. 8873-8892. 1995

Guil, J. M., Guil-Lopez, R., Perdigon-Melon, J.A., Corma, A., Determining the topology of zeolites by adsorption microcalorimetry of organic molecule, *Microporous Mesopourous Materials*, vol. 22, pp. 269–279, 1998.

Guo, T., Bai, Z., Wu, C., Zhu, T., Influence of relative humidity on the photocatalytic oxidation (PCO) of toluene by TiO2 loaded on activated carbon fibers: PCO rate and intermediates accumulation, *Applied Catalysis B: Environmental*, vol. 79, pp. 171-178, 2008.

Hermia, J., Vigneron, S., Catalytic incineration for odor abatement and VOC destruction. *Catalysis Today*. vol.17, pp. 349-358, 1993.

Hirota, K., Sakai, H., Washio, M., Kojima, T., Application of electron beams for the treatment of VOC streams. *Industrial Engineering Chemistry Research*, vol. 43, pp. 1185-1191, 2004

Horsley, J. A. , *Catalytica Environmental Report No E4*, Catalytica Studies Division, Mountain View, CA, USA, 1993.

Houari, A., Hamdi, B., Brendle, J., Bouras, O., Bollinger, J.C., Baudu, M., Dynamic sorption of ionizable organic compounds (IOCs) and xylene from water using geomaterial-modified montmorillonite, *Journal of Hazardous Materials*, vol. 147, pp. 738–745, 2007.

Huang, H., Li, W., Destruction of toluene by ozone-enhanced photocatalysis: Performance and mechanism, *Applied Catalysis B: Environmental*, vol. 102, pp. 449-453, 2011

Huang, Q., Vinh-Thang, H., Melekian, A., Eic, M., Trong-On, D., Kaliaguine, S., Adsorption of n-heptane, toluene and o-xylene on mesoporous UL-ZSM5 materials, *Microporous Mesoporous Matter*. 87 (2006) 224–234.

Huang, S., Zhang, C., He, H., Shaoyongn, In situ adsorption–catalysis system for the removal of o-xylene over an activated carbon supported Pd catalyst, *Journal of Environmental Sciences*, vol. 21, pp. 985–990, 2009.

Huang, Z., Kang, F., Liang, K., Hao, J., Breakthrough of methyethylketone and benzene vapors in activated carbon fiber beds, *Journal of Hazardous Materials*, vol. 98, pp. 107–115, 2003.

Hunter, P., Oyama, S.T., *Control of Volatile Organic Compound Emissions: Conventional and Emerging Technologies*, John Wiley & Sons, Inc., New York, 2000.

Hwang, K. S., Choi, D.K., Gong, S.Y., Cho, S.Y., Adsorption and thermal regeneration of methylene chloride vapor on an activated carbon bed, *Chemical Engineering Science, vol.* 52, pp. 1111–1123, 1997.

Jee, J. G., Lee, S.J., Lee, C.H., Comparison of the Adsorption Dynamics of Air on Zeolite 5A and Carbon Molecular Sieve Beds. *Korean Journal Chemical Engineering*, vol. 21, pp. 1183-1192, 2004.

Kansal, A., Sources and reactivity of NMHCs and VOCs in the atmosphere: A review. *Journal of Hazardous Materials*, vol 166, pp. 17-26, 2009.

Khan, F. I., Ghoshal, A. Kr., Removal of volatile organic compounds from polluted air, *Journal of Loss Prevention in the Process Industries*, vol. 13, pp. 527 – 545, 2000.

Kumar, T.P., Rahul, M., Kumar, A., Chandrajit, B., Biofiltration of Volatile Organic Compounds (VOCs)-An overview. *Research Journal of Chemical Sciences*, vol.1, pp. 83-92, 2011.

Kwong, C. W., Chao, C. Y. H., Hui, K. S., Wan, M. P., Catalytic Ozonation of Toluene Using Zeolite and MCM-41 Materials, *Environment Science Technology.*, vol. 42, pp. 8504-8509, 2008

Langmuir, I., The constitution and fundamental properties of solids and liquids. Part 1. Solids, *Journal of the American Chemistry. Society, vol.* 38 (11), pp. 2221, 1916

Le Cloirec, P., *Les Composés Organiques Volatils (COV) dans L'Environnement, TEC & DOC*, Paris, 1998

Le Cloirec, P., Fanlo, J.L., Gracian, C., Traitement des odeurs : les procédés curatifs, *Les techniques de l'ingénieur Traité environnement* (G 2971): pp. 1-14, 2003

Le Cloirec, P., Introduction aux traitements de l'air. *Techniques de l'Ingénieur,* G1700 v2 2008.

Lee, J. J., Kim, M. K., Lee, D.G., Kim, M.J., Ahn, H., Lee, C.H., Heat-exchange pressure swing adsorption process for hydrogen separation, *AIChE Journal,* vol. 54 pp. 2054–2064, 2008.

Lee, S. J., Hwan, J.J., Moon, J.H., Jee, J.G., Lee, C.H., Parametric study of the three-bed PVSA process for high purity O2 generation from ambient air, *Industrial Engineering Chemistry Research,* vol. 46, pp. 3720–3728, 2007.

Liu, J., Dong, M., Sun, Z., feng Qin, Z., Wanga, J., Sorption of xylenes in CoAlPO4-5 molecular sieves, *Colloids and Surface. A: Physicochemical and Engineering Aspects*, vol. 247, pp. 41–45, 2004.

Liu, P., Longa, C., Lia, Q., Qian, H., Lia, A., Zhang, Q., Adsorption of trichloroethylene and benzene vapors onto hyper-crosslinked polymeric resin, *Journal of Hazardous Materials*. Vol. 166, pp. 46–51, 2009

Long, C., Li, Y., Yu, W.H., Li, A.M., Removal of benzene and methyl ethyl ketone vapor: comparison of hyper-crosslinked polymeric adsorbent with activated carbon, *Journal of Hazardous Materials*. Vol. 203–204, PP. 251–256, 2012.

Makowski, W., Kustrowski, P., Probing pore structure of microporous and mesoporous molecular sieves by quasi-equilibrated temperature programmed desorption and adsorption of n-nonane, *Microporous Mesoporous Materials,* vol. 102, pp. 283–289, 2007.

Manero, M. H., Roustan, M., Etude comparative de divers procédés de traitement des effluents gazeux à faible concentration, *L'eau, l'industrie, les nuisances,* vol.184, pp. 85-88, 1995

Manjare, S. D., Ghoshal, A.K., Studies on adsorption of ethylacetate vapor on activated carbon, *Industrial Engineering Chemistry Research,* vol. 45, pp. 6563–6569, 2006.

Martin, A.M., Nolen, S.L., Gess, P.S., Baesen, T.A., Control Odors from CPI Facilities, *Chemical Engineering Progress*, vol 88, pp. 53-61, 1992.

Meininghaus, C. K. W., Prins, R., Sorption of volatile organic compounds on hydrophobic zeolites, *Microporous Mesopourous Materials*, vol. 35–36, pp. 349–365, 2000.

Middelton, P., Singh, H.B., Nriager, J., (Eds.), *Sources of Air Pollutants in Composition, Chemistry and Climate of Atmosphere*, John Wiley and Sons, New York, 1995.

Molina, M., Rowland, F.S., Stratospheric sink for chlorofluoromethanes: chlorine atom-catalyzed destruction of ozone, *Nature* vol. 249, pp. 810–812, 1974

Monneyron, P., *Procédés hybrides adsorption/oxydation en phase gazeuse-Application au traitement des composés organiques volatils*, PhD Thesis, INSA Toulouse, 2004.

Moretti, E. C., Mukhopadhyay, N., VOC control: current practices and future trends. *Chemical Engineering Progress*, vol. 89, pp. 20-26, 1993

Muzenda, E. , *Pre-treatment Methods in the Abatement of Volatile Organic Compounds: A Discussion, International Conference on Nanotechnology and Chemical Engineering* (ICNCS'2012), Bangkok, December 21-22, 2012

N. R. C. *The Scientific basis for estimating emissions from animal feeding operations.* National Academy Press, Washington, DC, 2002

Nathanson, T., Indoor Air Quality in Office Buildings: A Technical Guide. *Health Canada*, Ottawa, Ontario. 93-EHD-166, 1995

Naydenov, A., Mehandjiev, D., Complete oxidation of benzene on manganese dioxide by ozone. *Applied Catalysis A: General*, vol. 97, pp. 17-22, 1993.

Otero, M., Zabkova, M., Grande, C.A., Rodrigues, A.E., Fixed bed adsorption of salicylic acid onto polymeric adsorbents and activated charcoal, *Industrial Engineering Chemistry Research, vol.* 44 , pp. 927–936, 2005.

Paints Directive 2004/42/EC, Directive 2004/42/CE of the European Parliament and of the Council of April 21, 2004 on the limitation of emissions of volatile organic compounds due to the use of organic solvents in certain paints and varnishes and vehicle refinishing products amending Directive 1999/13/EC, *Official J. Eur. Union L.* 143, 87– 96.

Parmar, G. R., Rao, N. N., Emerging Control Technologies for Volatile Organic Compounds, *Environmental Science and Technology*, vol 39, pp. 41-78, 2008.

Pennell, K. D., Rhue, R.D., Rao, P.S.C., Johnston, C.T., Vapor-phase sorption of p-xylene and water on soils and clay minerals, *Environment Science Technology*, vol. 26, pp. 756–763, 1992.

Pilverdier, E., *Apport de la Calorimétrie 'a la Connaissance des Interactions Zéolithe-Adsorbat et Corrélations Avec les Données Structurales: Cas des Systèmes Benzène et Dérivés Méthyles du Benzène/Zéolithe type Faujasite.* PhD Thesis, University of Bourgogne, Dijon, France, 1995.

Qu, F., Zhu, L., Yang, K., Adsorption behaviors of volatile organic compounds (VOCs) on porous clay heterostructures (PCH), *Journal of Hazardous Materials*, vol. 170, pp. 7–12, 2009

Rafson, J. H., *Odor and VOC Control Handbook.* "Section 8 -Thermal Oxidation." McGraw Hill. New York, NY. Pp. 8.31 - 8.65, 1998

Residual Risk, Report to Congress, United States Environmental Protection Agency, North Carolina, Research Triangle Park, pp. 81- 84, 1999.

Rusu, A. O., Dumitriu, E., destruction of volatile organic compounds by catalytic oxidation, *Environmental Engineering and Management Journal*, vol. 2 (4), pp. 273-302, 2003.

Ruthven, D. M., *Principles of Adsorption Processes*. New York: John Wiley, 1984

Sahu, K. L., Volatile organic compounds and their measurements in the troposphere, *current Science*, vol 102 (12), pp.1645-1649, 2012.

Schork, J. M., Fair, J.R., Parametric analysis of thermal regeneration of adsorption beds, *Industrial Engineering Chemistry Research, vol.* 27, pp. 457–469, 1988.

Sheintuch, M., Matatov-Meytal, Y.I., Comparison of catalytic processes with other regeneration methods of activated carbon, *Catalysis Today,* vol. 53, pp. 73–80, 1999.

Shim, W. G., Lee, J.W., Moon, H., Adsorption equilibrium and column dynamics of VOCs on MCM-48 depending on pelletizing pressure, *Microporous and Mesoporous Materials*, vol 88, pp. 112–125, 2006

Simonot-Grange, M. H., Bertrand, O., Pilverdier, E., Bellat, J.P., Paulin, C., Differential calorimetric enthalpies of adsorption of p-xylene and m-xylene on Y faujasites at 25∘C, *Journal of Thermal Analysis,* vol. 48, pp. 741–754, 1997.

Singh, H. B., Zimmerman, P., Atmospheric distributions and sources of non-methane hydrocarbons, in: J.O. Nriagu (Ed.), *Gaseous Pollutants: Characterization and Cycling*, Wiley, New York, 1992.

Sleiman, M., Conchon, P., Ferronato, C., Chovelon, J.-M., Photocatalytic oxidation of toluene at indoor air levels (ppbv): Towards a better assessment of conversion, reaction intermediates and mineralization. *Applied Catalysis B: Environmental*, vol. 86, pp. 159-165, 2009

Song, L., Rees, L.V.C., Adsorption and diffusion of cyclic hydrocarbon in MFI-type zeolites studied by gravimetric and frequency-response techniques, *Microporous Mesopourous Materials*, vol. 35–36, pp.301–314, 2000.

Talapatra, A., Srivastava, A., Ambient Air Non-Methane Volatile Organic Compound (NMVOC) Study Initiatives in India – A Review. *Journal of Environmental Protection*, vol 2(1), pp. 21-36, 2011.

Talu, O., Guo, C.J., Hayhurst, D.T., Heterogeneous adsorption equilibria with comparable molecule and pore sizes, *J. Phys. Chem.* 93 (1989) 7294–7298.

Thalhammer, H., VOC and odour removal in applications with dust loaded flue gases. *Odours and VOCs Journal.* Vol 1,10, pp. 219, 1995.

Tichenor, B. A., Palazzolo. M.A., Destruction of volatile organic compounds via catalytic incineration, *Environmental Progress*, vol. 6 pp. 172-176, 1987.

Torrents, A., Damera, R., Hao, O.J., Low-temperature thermal desorption of aromatic compounds from activated carbon. *Journal of Hazardous Materials,* vol. 54 (3), pp. 141–153, 1997

US. EPA, Sources of Indoor Air Pollution – Organic Gases, *http://www.epa.gov*, 2003.

US. EPA, *Off-gas treatment technologies for soil vapor extraction systems: state of the practice,* USEPA, 128 p, (2006).

US. EPA, *EPA's Terms of Environment Glossary, Abbreviations, and Acronyms.* Code of Federal Regulations, Title 40: Protection of Environment. Chapter 1, Environmental Protection Agency Subchapter C, Part 51, Subpart F, 51100, Retrieved on 2009-02-08.

US. EPA, An Introduction to Indoor Air Quality (IAQ) – Volatile Organic Compounds (VOCs). *http://www.epa.gov/iaq/voc2.html*, 2011.

US. EPA. Volatile organic compounds (VOCs). Retrieved 17/8/2012 from *http://www.epa.gov*, 2012.

Valdés, H., Solar, V., Cabrera, E.H., Veloso, A.F., Zaror, C.A., Control of released volatile organic compounds from industrial facilities using natural and acid-treated mordenites: the role of acidic surface sites on the adsorption mechanism, *Chemical. Engineering Journal.* 244 (2014), 117-127.

Van der Vaar, D. R., Marchand, E.G., Bagely-Pride, A., Thermal and catalytic incineration of volatile organic compounds. Critical Reviews in *Environmental Science and Technology.* Vol. 23, pp. 203-236, 1994.

VOC Solvents Directive 1999/13/EC, *Council Directive 1999/13/EC of 11 March 1999 on the limitation of emissions of volatile organic compounds due to the use of organic solvents in certain activities and installations.*

Volkamer, R., Sheehy, P., Molina, L.T., Molina, M.J., Oxidative capacity of the Mexico City atmosphere - Part 1: A radical source perspective. *Atmospheric Chemistry and Physics*, vol. 10, pp. 6969-6991, 2010.

Wang, C. M., Chang, K.S., Chung, T.W., Adsorption equilibria of aromatic compounds on activated carbon, silica gel, and 13X zeolite, *Journal of Chemical Engineering Data*, 49 (2004) 527–531.

Wang, X., Daniels, R., Baker, R.W., Recovery of VOCs from high-volume, low -VOC-concentration air streams. *AIChE Journal*, vol 47, pp. 1094-1100, 2001.

World Health Organization, 1989. *"Indoor air quality: organic pollutants."* Report on a WHO Meeting, Berlin, 23-27 August 1987. EURO Reports and Studies 111. Copenhagen, World Health Organization Regional Office for Europe.

Wu, C., Chung, T., Yang, T.C.K. Chen, M., Dynamic determination of the concentration of volatile alcohols in a fixed bed of zeolite 13X by FT-IR, *Journal of Hazardous Materials.* vol. B137, pp. 893–898, 2006.

Xi, Y., Reed, C., Lee, Y.-K., Oyama, S. T., Acetone Oxidation Using Ozone on Manganese Oxide Catalysts. *Journal of Physical Chemistry B,* vol. 109, pp. 17587–17596, 2005.

Yang, R. T., *Gas separation by adsorption processes*, Butterworths, Boston, 1987.

Yoon, Y. H., Nelson, J.H., Application of gas adsorption kinetics. I. A theoretical model for respirator cartridge service life, *American Industrial Hygiene Association Journal.* 45 (8) (1984) 509–516.

Yu, F. D., Luo, L., Grevillot, G., Electrothermal swing adsorption of toluene on an activated carbon monolith. Experiments and parametric theoretical study, *Chemical Engineering and Processing*, vol. 46, pp. 70-81, 2007

Yun, J. H, Choi, D. K., Kim, S. H., Adsorption of organic solvent vapors on hydrophobic Y-Type zeolite. *AIChE Journal*,;vol.44, pp. 1344–1350, 1998

Yun, J. H., Choi, D.K., Moon, H., Benzene adsorption and hot purge regeneration in activated carbon beds, *Chemical Engineering Science.* Vol. 55, pp. 5857–5872, 2000.

Zaitan, H., Bianchi, D., Achak, O., Chafik, T., A comparative study of the adsorption and desorption of o-xylene onto bentonite clay and alumina, *Journal of Hazardous Materials*, vol. 153, pp. 852–859, 2008.

Zaitan, H., Chafik, T., FTIR determination of adsorption characteristics for volatile organic compounds removal on diatomite mineral compared to commercial silica, *Comptes rendus Chimie*, vol. 8 (9–10), pp. 1701–1708, 2005.

Zaitan, H., Feronnato, C., Bianchi, D., Achak, O., Chafik, T., Study of textural and adsorptive properties of a Moroccan diatomite: application to the treatment of air polluted with a volatile organic compound, *Annales de Chimie Sciences des Matériaux*, vol. 31 (2) pp. 183–196, 2006.

Zaitan, H., Korrir, A., Chafik, T., Bianchi, D., Evaluation of the potential of volatile organic compound (di-methyl benzene) removal using adsorption on natural minerals compared to commercial oxides, *Journal of Hazardous Materials*, vol. 262, pp.365– 376, 2013.

Zhao, X. S., Ma, Q., Lu, G. Q., VOC Removal: Comparison of MCM-41 with Hydrophobic Zeolites and Activated Carbon. *Energy & Fuels,* vol. 12, pp.1051-1054, 1998

Zhao, Z., Li, X., Li, Z., Adsorption equilibrium and kinetics of p-xylene on chromium-based metal organic framework MIL-101, *Journal of Chemical Engineering*, vol.173, pp. 150–157, 2011.

Zuo, S., Liu, F., Zhou, R., Qi, C., Adsorption/desorption and catalytic oxidation of VOCs on montmorillonite and pillared clays, *Catalysis Communications*, vol. 22, pp. 1–5, 2012. Source CITEPA, Format SECTEN – 2013

INDEX

A

abatement, xi, 145, 146, 163, 165, 170, 180, 189, 191, 202, 205

ability, x, 8, 104, 107, 130, 153, 177, 191

absorbent, 182

absorbers, 181

absorption, xi, 169, 171, 181

abundant, x, 103, 133, 184

acceleration, 9, 19

accumulation, 42, 74, 132, 205

acetaldehyde, 146, 165

acetic acid, 161, 167

acidification, 133

actions, 29, 126, 129, 180

activation, 150

active, x, xi, 91, 107, 115, 144, 145, 147, 149, 151, 152, 153, 156, 158, 161, 162, 164, 178, 195

activity, xi, 85, 139, 144, 146, 149, 151-154, 156, 158, 161, 162, 164-167, 176, 178

adaptations, 180

adhesive agents, 176

adhesives, 105

admissions, 113

adsorbents, xi, 169, 170, 183, 184, 191, 192, 194, 195, 196, 203, 204

adsorber flow, 184

adsorption, xi, 78, 169, 171, 181, 182, 183, 184, 185, 191, 192, 193, 194, 195, 196, 197, 199-207

adult species, 119, 120, 121

advantages, 4, 186, 191

advection, 20, 22

aerodynamic, viii, 2, 3, 8

aerodynamic loads, 8

aerodynamic properties, 3, 8

aerosols, 36, 80, 98, 112, 130, 133

agents, 105, 147, 173, 176

agglomeration, 147, 148, 156

aggregation, 147

agricultural, 80, 91, 138, 184

air masses, 16, 18, 19, 20, 21, 22, 23, 81

air pollutant, 80, 101, 125, 170, 186, 189, 190

air pollution, ix, x, 43, 75, 79, 80, 113, 134, 141, 143, 145, 170, 177, 188

air quality, ix, 26, 38, 80, 81, 87, 93, 129, 130, 139, 140, 170, 174, 206

air quality models, 87, 93

aircrafts, 51

alcohols, xi, 28, 55, 94, 144, 145, 173, 176, 184, 186, 199

aldehydes, 28, 145, 173, 186

algorithm, 82

alkanes, xi, 31, 33, 46, 87, 94, 144, 145, 146, 154, 162, 165, 176, 206

alkenes, 33, 85, 87, 94, 95, 146

alterations, 28, 113, 135

ambient concentrations, 85, 141
ambient conditions, 44, 92, 100
ambient measurements, 85, 87
ambient pressure, 183
ambient temperature, 36, 43, 62, 70, 195
ammonia synthesis, 145, 164, 166
ammonium carbonate, 171
amounts, 42, 43, 47, 48, 54, 59, 104, 108, 125, 174
anthropic origin, 42
anthropogenic, vii, ix, 7, 25, 79, 80, 83, 87, 104, 105, 109, 138, 173, 174
Anthropogenic alkenes, 85
anthropogenic VOCs, ix, 80, 81, 83
application, 52, 53, 91, 96, 145, 154, 162, 166, 182, 200
approaches, x, 47, 104, 126
approximation, 126
areas, ix, 8, 32, 33, 36, 80, 81, 82, 83, 88, 89, 91, 93, 94, 126, 129, 133, 173, 174, 176, 183, 192
aromatic hydrocarbons, 28, 43, 172
aromatics, 87, 145, 146, 173, 176, 198
aspects, 115, 136, 139
assignation, 129
associations, 113, 126
assurance, 87
atmospheric background concentration, 4
atmospheric boundary layer, 3, 11, 12, 13, 23, 24, 26
atmospheric chemistry, 105, 112, 130, 133, 135
atmospheric evolution, 92
atmospheric flows, 7
atmospheric measurement, 82, 87, 93
atmospheric photochemical processes, 81
atmospheric pollution, vii, 42
atmospheric reactivities, ix, 80
atomic oxygen, 108, 178
atoms, 178
autumn, x, 38, 90, 93, 104, 115, 116, 124, 125
average, 4, 6, 8, 17, 20, 21, 32, 33, 36, 44, 59, 67, 83, 84, 85, 86, 93, 113, 126, 140, 149, 190, 192, 196

AVOC emission, 81

B

beads, 185
beam expander, 9
beam separation, 9
behavior, 42, 48, 116, 126
bell-shaped distribution, 86, 89
benefits, x, 103, 183, 197
bentonite, 78, 184, 191, 192, 195-197, 201
bentonite clay, 78, 196, 197, 201
benzene, 31, 33, 83, 177, 198, 199, 200, 204, 207
benzene derivatives, 33
bioadsorbents, 184
biodegradable, 186
biodegradation, 181
biofilter, 186
biofiltration, xi, 169, 186, 200
Biogenic, v, ix, x, 1, 41, 42, 76, 79, 80, 83, 87, 96, 97, 114, 126, 133, 135, 136, 137, 140, 141
biogenic emissions, x, 7, 23, 24, 77, 83, 86, 87, 104, 132
biogenic hydrocarbons, 94
biogenic sources, ix, 79, 83, 85
biogeochemistry, 131
biological treatment, 171, 186
biomass, 29, 36, 38, 76, 90, 126, 176
bioreactors, 187
bioscrubber, 186
biosolids, 187
biosynthesis, 106, 107
biosynthetic, 106, 107
biradical, 111
boiling point, 53, 62, 70, 72, 147, 172, 184
boiling temperature, 51, 65
boundary conditions, 3, 23
boundary layer, viii, 2, 3, 7, 8, 9, 10, 11, 12, 13, 19, 23, 24, 26
boundary layer thickness, 7, 9
BVOC abundance, 82, 84
BVOC flux, 91
BVOCs changes, 52

Index 211

BVOCs composition, 51, 63, 70
BVOCs concentration, 74
BVOCs emission, ix, x, 42, 43, 44, 47, 48,
    49, 50, 51, 52, 53, 55, 62, 64, 65, 68, 69,
    74, 104, 105
BVOCs mixture, 57, 59, 62, 63, 69, 70
BVOCs molecules, 52
BVOCs production, 62

C

calcined, 154, 158
calculation, 115, 132
calibration, 11, 47, 72
candidate, 154, 196
canister samples, 82
canisters, 87, 98
canopy, vii, viii, 1, 2, 3, 4, 6, 8, 12, 13, 15,
    16, 19, 21, 22, 23, 24, 25, 131
canopy flow, 3
capacity, 43, 91, 114, 125, 130, 138, 171,
    182, 183, 192, 193, 195, 196, 205
capture, x, 82, 103, 171, 181
carbon cycle, 28, 112
carbon dioxide, x, 28, 38, 43, 103, 164, 171,
    185, 186, 190
carbon monoxide, xi, 28, 38, 97, 112, 144,
    164, 171
carbonates, 171
carbons, 131, 183, 184, 188, 190, 191, 193,
    195, 196
carboxylic acids, 28
carrier gas, 31, 46
cartography, 8
catalysis, xi, 146, 163, 166, 169, 186, 189,
    205
catalysts, x, xi, 144, 145, 146, 147, 149,
    151, 152, 153, 154, 155, 156, 157, 158,
    159, 160, 161, 162, 163, 164, 165, 166,
    167, 189, 200
catalytic oxidation, 144, 146, 150, 156, 163,
    171, 181, 186, 190, 198, 205
catalytic ozonation, 190, 198
cedars, 94
century, 114

challenge, 171
characteristics, viii, 2, 25, 26, 58, 62, 68, 75,
    84, 85, 90, 97, 101, 105, 126, 130, 188,
    190, 201
characterization, ix, 29, 74, 76, 80, 94, 147,
    152, 165, 166
chemical, vii, ix, x, 1, 51, 54, 56, 59, 60, 62,
    66, 70, 72, 79, 80, 82, 87, 93, 98, 103,
    108, 109, 110, 111, 112, 114, 116, 126,
    129, 153, 170, 171, 172, 173, 176, 185,
    190
chemical families, 54, 62, 70, 176
chemical fire retardant, 51
chemical formulas, 54, 59
chemical reactions, vii, ix, 1, 79, 80, 109
chemical retardant, 51
chemisorption, 149
chemisorptive, 150
chemistry, viii, ix, 2, 38, 42, 80, 82, 92, 95,
    105, 111, 112, 130, 133, 134, 135, 136,
    137, 141, 145, 146, 163, 164, 166, 200
chlorine, 178, 203
cholinesterase, 177
chromatography analysis, 87
chromatography technique, 4
circumstances, 74, 126
civilization, 191
classification, 4, 6, 33, 171, 172
classification scheme, 4, 6
clay, 51, 76, 78, 191, 196, 197, 200, 201,
    202
cleaners, vii, 170
climate, 28, 81, 95, 112, 114, 125, 139, 141,
    170
climate conditions, 81
clusters, 152, 154, 162
coalescence, 147
coaxial, 185
co-existence, 152, 162
colloid, 146, 147, 148, 166
combination, 82, 171, 178
combustion, viii, ix, xi, 10, 27, 28, 29, 30,
    34, 36, 37, 42, 43, 74, 75, 76, 144, 145,
    156, 158, 161, 162, 163, 164, 165, 167,
    173, 181, 185, 186

combustion chamber, viii, 10, 27, 29, 30, 186

combustion efficiency, 34

commercial, 31, 46, 105, 162, 192, 195, 196, 197, 200, 201

communes, 126, 129

community, 94, 129

compilation, 11

complex, 12, 23, 24, 81, 130

components, 31, 39, 46, 51, 59, 89

composition, 28, 37, 43, 51, 57, 59, 61, 62, 63, 65, 68, 70, 76, 81, 89, 90, 95, 119, 136, 138, 172

compound, ix, 30, 41, 46, 48, 51, 55, 57, 59, 62, 65, 67, 70, 71, 72, 80, 99, 100, 106, 112, 116, 126, 130, 132, 136, 139, 161, 171, 172, 183, 197, 200, 201

compressors, 182

concentration, viii, 2, 3, 4, 10, 13, 14, 17, 20, 55, 74, 83, 85, 95, 106, 109, 112, 113, 151, 171, 180, 182, 183, 186, 187, 188, 190, 191, 199, 203, 206

condensation, 171, 181, 182, 184, 200

condenser, 182

coniferous, 76, 94, 99, 137

constituents, 38, 51, 62, 70, 105

consumption, 30, 38, 80, 109, 185

contaminant, 130, 183

contamination, 152

contribution, ix, x, 3, 80, 83, 85, 87, 88, 89, 90, 94, 95, 98, 104, 130, 133, 138, 144, 145

contributor, 81, 83, 84, 85, 88, 90, 93

convection, 19, 43

conversion, 151, 152, 153, 156, 157, 159, 203

co-precipitation, 154

Coriolis forces, 7

corrosive, 185

Corsica, v, viii, 27, 28, 29, 30, 32, 33, 34, 35, 36

critical conditions, 32, 36

cross-validation, 92

cryogenic, 182, 203

crystalline, 150, 153, 157

crystallites, 151, 152, 157, 162

crystallization, 150

cyclic, 105, 203

cylinders, 4, 185

D

damages, 28

database, ix, 29, 42, 44, 74, 82, 88, 90, 91, 92, 101

Deacon process, 145

decomposition, 104, 111, 190

decontamination, x, 103, 104

deficient, 129, 151

deflection, 18, 19, 20, 21

degradation, 42, 47, 51, 53, 62, 65, 70, 74, 96, 111, 140, 184, 189

degreasers, vii, 170, 176

dendrimers, 147

dense, 4, 12, 13, 130

density, 9, 26, 75, 91, 92, 196

dependence, 93, 108

depletion, vii, 87, 170, 177, 178

deposition, 148, 149

deposits, 104

desorption, xi, 46, 78, 169, 185, 195, 198, 200, 201, 205, 206

destruction, xi, 169, 178, 181, 188, 198, 202, 203

detection, 20

detergents, vii, 170

deterioration, 161

deviations, 6, 13, 15, 35, 122, 126

diatomite, 184, 191, 197, 200, 201

diesel, xi, 144, 176

differences, 14, 58, 61, 62, 68, 70, 92, 116, 125, 129, 132

diffraction, 145, 148, 154

diffusion, 4, 6, 26, 106, 153, 198, 202, 203

diffusion class, 4, 6

dimensions, 7, 157

dioxins, 185

direct emissions, 36

direction, 4, 6, 7, 12, 14, 92

disadvantage, 186

Index

213

dispersion, vii, 1, 3, 4, 6, 7, 14, 16, 20, 23, 25, 26, 147, 149, 152, 154, 159, 161, 162

dispersion process, vii, 1, 3, 4, 6, 7, 14, 23, 26

displacement height, 11

disruption, 112

dissociation, 178, 190

dissolution, 181

distances, 21

distributions, 13, 16, 18, 89, 91, 148, 200

diurnal variations, 84, 90, 100

diversity, 92, 94

divisions, 106

dolomite, 184

double bonds, 111

drag coefficient, 3

drainage, 129

driest seasons, 32, 35, 36

durability, 162

E

ecological, x, 28, 103, 126

economic development, 80

economic region, ix, 79

ecosystems, 28, 105, 139

effects, vii, xi, 7, 25, 28, 34, 38, 42, 43, 44, 47, 51, 75, 97, 98, 105, 113, 114, 129, 130, 134, 138, 139, 141, 142, 166, 167, 169, 170, 171, 177, 179

efficiency, 34, 171, 182, 186, 188, 190, 191, 196

efficient, 58, 176, 186, 188, 189, 195

effluent flow, 182

effluents, 182, 183, 196, 203

electrocatalysts, 145

electrochemistry, 145

electrodes, 31, 145

electron, 31, 47, 144, 145, 147, 148, 202

electronic, 173

electrostatic, 180

emergency, 113

Emission and Chemical transformation of Organic compounds (ECHO), viii, 1

emission data, 89

emission factor, 30, 39, 57, 89, 90, 92, 99, 104, 108, 117, 118, 119, 120, 122, 124, 138

emission fluxes, 87, 88, 138

emission input, 89

emission intensity, 89

emission inventory, x, 82, 87, 88, 89, 92, 100, 101, 104, 127, 180

emission peak, 65, 72

emission rates, 2, 77, 88, 90, 91, 92, 98, 99, 100

empirical laws, 11

enclosure, 43, 44, 45, 51, 52, 58, 62, 74, 115, 131, 132

energy consumption, 80, 185

energy ionization, 31

environment, vii, viii, ix, x, xi, 2, 25, 42, 43, 44, 58, 78, 79, 80, 81, 104, 130, 169, 170, 171, 177, 180

environmental, vii, x, xi, 7, 104, 106, 107, 114, 116, 125, 126, 129, 130, 138, 169, 170, 177, 186, 200

equilibrium, 51, 62, 70, 191, 195, 206, 207

escape, 180

esters, 28, 33, 94, 173

ethane, 10, 11, 16, 20

ethyl acetate, 146, 161, 165

evaporation, 51, 53, 62, 70

evaporation process, 51, 62, 70

evolution, 24, 48, 51, 59, 92

exacerbation, 80, 113

exchangers, 182

exothermicity, 162

exotic, x, 104, 115, 116, 117, 118, 119, 120, 121, 122, 123, 124, 125, 127, 129, 130

exotic species, 115, 117, 119, 121, 123, 129

exotic trees, x, 104, 119, 130

experiment, 4, 6, 39, 137

experimental device, 29, 30

experimental protocol, 45, 58

explosivity, 191

exposition, 121

extensive, 7, 23, 28, 32, 90, 183

extraction methods, 58

F

field data, viii, 2, 3, 11, 12, 24, 91
field experiments, viii, 1, 3, 6, 13, 23
fight, 51
fire data, 29
fire extinguishers, 178
fire regimes, 28
fire retardants, 51, 75, 172, 178
fires, viii, ix, 27, 28, 29, 30, 32, 33, 34, 35,
    36, 37, 38, 39, 42, 43, 51, 74, 75, 78, 174
firs, 94
flame ionization detector, 31
flammability, 77, 190
flammable, 42, 184
flashpoint, 42
floods, 129
flow, viii, 2, 3, 4, 7, 8, 9, 11, 12, 13, 16, 18,
    21, 23, 25, 26, 45, 46, 154, 180, 182,
    184, 187, 188, 190, 195
flow acceleration, 19
Flow Ionisation Detectors, 16
flow measurements, 12
flow properties, 3, 23
flow rate, 4, 11, 180, 184, 187, 190
flow resistor, 4
flow velocity, 7, 9
flower fragrances, 105
flue gases, 180, 200
fluid, 7, 9, 26, 51
focal distance, 9
foliar density, 91, 92
footprint, viii, 2, 3, 16, 25
footprint experiments, viii, 2, 3, 16
forest area, viii, 2, 3, 7, 9, 10, 11, 12, 15, 16,
    18, 20, 23, 24, 32, 36, 74, 94
forest fetch, 4, 7
forest fires, ix, 28, 35, 37, 38, 39, 42, 43, 51,
    74, 75, 78, 174
forest stand, viii, 2, 3, 11, 43
formaldehyde, 93, 184
formation, vii, ix, 80, 81, 82, 83, 84, 85, 86,
    87, 88, 93, 95, 96, 98, 99, 100, 107, 108,
    112, 114, 129, 130, 137, 140, 141, 150,
    151, 153, 156, 158, 161, 170, 177, 178,
    180, 185, 189
formulations, xi, 144
fractions, 16
France, viii, 1, 27, 28, 29, 41, 76, 143, 174,
    175, 176, 197, 199
frequency, 10, 43, 203
friction velocity, 11
fringes, 148, 157
fruit trees, 113
fuel, viii, 27, 28, 29, 30, 34, 75, 173
furans, 28, 33, 185

G

gardens, 105
gas chromatograph, 4, 5, 36, 47, 76, 78, 87
gas exchange, 121
gas flow, 180, 184
gaseous, xi, 100, 126, 169, 172, 180, 182,
    183, 184, 186, 189, 190, 191, 196, 202
gases, viii, ix, x, 2, 11, 28, 29, 36, 37, 38,
    41, 43, 60, 74, 103, 105, 180, 185, 188,
    200
generation, 92, 191, 205
geographical area, 87
geological hydrocarbon, 104
geometric lengths, 7
geometric scale, 3, 7
geometry, 3, 16, 28
global change, 75, 114, 139, 141
global warming, 37, 43
glycols, 147
goals, 126
gradient, 4, 6, 8, 9, 12, 13, 78
grassland, 39, 91
greenhouse, 28, 43, 58, 105, 108, 170, 177
greenhouse gases, 28, 105
grids, 91
ground, 4, 6, 9, 11, 12, 13, 14, 15, 16, 18,
    19, 21, 22, 24, 91, 177
ground area, 16, 21, 22
ground data, 91
growth, x, 9, 104, 108, 113, 115, 121, 140,
    188

Index   215

H

halogenated, 145, 178, 186, 190
headspace solid phase microextraction, 58, 76
heat, 44, 50, 52, 54, 55, 125, 183, 185, 186, 190, 195
heat flux, 44, 50, 52, 54, 55
heating, 47, 48, 49, 50, 74, 147, 184
heating rate, 47, 48, 49, 74
heating time, 47, 48, 49, 74
height, 4, 6, 7, 8, 9, 10, 11, 12, 16, 20, 44, 50, 92
helium, 31, 46, 47
herbivore, 105
hermetic enclosure, 43, 44, 45, 74
heterogeneous, 150, 163, 203
histograms, 149
homogeneous, 16
Hong Kong, 79, 80, 81, 82, 83, 85, 86, 89, 90, 91, 92, 93, 94, 95, 97, 98, 99, 100, 101, 138
hormones, 105
Humidity, 140
Hydric stress, 108
hydrocarbons, 28, 33, 42, 51, 63, 70, 81, 94, 100, 101, 105, 109, 110, 132, 141, 142, 145, 146, 172, 173, 174, 190, 200
hydrodistillation, 58, 76
hydrogen, xi, 144, 147, 152, 154, 162, 197, 202
hydrogenation, 145, 153, 164
hydrology, 131
hydroperoxy radicals, 109
hydroperoxyl, 112
hydrophilic, 184
hydrophobic, 183, 184, 199, 201, 205
hydroxyl radical, 108, 109, 110, 112, 140, 178

I

identification, viii, 27, 29, 31, 46, 125
imminent, 129

immobilization, 148
impacts, 81, 87, 92, 93, 95, 99, 170, 177
implementation, 180
impregnation, 147, 151, 154
incineration, 181, 185, 198, 199, 202
incinerators, 185
incorporation, 88
Index, vi, x, 4, 104, 115, 116, 117, 118, 120, 123, 124, 132, 209
industrial, xi, 75, 82, 85, 91, 105, 145, 146, 169, 170, 173, 174, 176, 180, 182, 196, 200
industries, vii, 170, 173, 184
inflammation, 113
influence, xi, 7, 8, 25, 26, 48, 51, 54, 68, 75, 76, 81, 88, 92, 99, 108, 138, 144, 152, 158, 162, 197
information, vii, viii, x, 1, 2, 3, 4, 23, 87, 89, 91, 92, 94, 104, 111, 115, 121, 130, 131, 133, 147, 180
inhomogeneous, 21
inner surfaces, 87
inorganic, 76, 147, 170, 188
input parameters, 89, 92
instrument, 126
integral length scales, 12
integration, x, 104
intensive, 93, 149, 189
interactions, 23, 131, 133, 136
intermediate, 111
interpolation, 31, 46
inventories, 89, 126, 139, 142
investments, 180
ion, 191
ion-exchange, 191
ionization, 31, 46, 47
ionization energy, 47
irregular, 157
irritation, 177
isomers, 193, 202, 204
isoprene abundance, 93
isoprene concentrations, 91
isoprene emissions, 91, 95, 98, 101, 133, 136, 141
isoprene flux, 91, 95

216 Index

isoprene levels, 83, 84, 86, 93
isoprene photochemical, 86
isoprenoid compounds, 28
isotherm, 31, 183, 201

K

ketones, 28, 145, 173, 186
key contributor, 83, 93
kinetics, 171, 207
knowledge, 29, 123

L

laboratory experiments, viii, 1, 2
laboratory measurements, 90, 132
landscape, 43, 91
large-scale flow distortion, 16
lateral dispersion, 16
layer profile, 8
leaf, 77, 94, 105, 106, 113, 126, 136, 142
Leaf Area Index (LAI), 3
levels, 80, 81, 82, 83, 84, 85, 86, 87, 88, 92,
    93, 94, 113, 114, 139, 170, 171, 177,
    180, 182, 184, 203
ligands, 147
light flux, 82
limitations, xi, 170, 186
limonene, ix, 41, 42, 43, 47, 50, 51, 53, 64,
    68, 70, 71, 74, 75, 106, 108, 109, 130,
    132
linear, 11, 31, 46, 105
liquid fuels, vii, 170
liquid phase, 4
liquids, 51, 201
load, 28, 34, 186
local flow distortions, 16
local geometry, 16
local porosity, 8
locations, 5, 12, 23, 81, 82, 83, 84, 93, 95
logarithmic law, 11
longitudinal integral length scale, 12
longitudinal pressure gradient, 9
longitudinal velocity, 13

loss of biodiver, 28
lubricants, vii, 170

M

magnetic, 147, 173
magnitude, viii, 2, 14, 23, 57, 88, 89, 113
major, ix, x, xi, 7, 42, 43, 48, 51, 54, 55, 57,
    59, 62, 65, 67, 70, 71, 72, 74, 79, 80, 81,
    83, 85, 86, 89, 94, 106, 111, 125, 126,
    143, 145, 162, 169, 177, 181
major compounds, 48, 54, 55, 57, 62, 74
management, 26, 32, 36, 121, 129, 140
mangroves, 91
manipulation, 108
manufacturing, 173, 174, 176
maquis, viii, 27, 28, 30, 32, 34
mass flow, 11, 13
mass flow controller, 11
material, 8, 44, 154, 183, 184, 188, 191
mathematical, 115
measurement, 4, 11, 12, 19, 23, 82, 87, 93,
    95
mechanism, 96, 111, 163, 190, 200
Mediterranean species, 44, 62, 69, 77
membrane separation, 182
membranes, 105, 181, 188
mesh, viii, 2, 3, 8, 10, 91
mesopore, 192
mesoporous, 163, 192, 204, 206
metabolites, 42, 138
metallic carbides, 171
metallic mesh, viii, 2, 3, 8, 10
metallic mesh rings, 8
metals, x, xi, 144, 145, 146, 154, 161, 163
meteorological conditions, 4, 14, 33, 36, 88,
    92
meteorological field data, 12
meteorological parameters, 82, 89
meteorological tower, 4, 5, 6, 12, 13, 14, 15,
    19, 21
methanation, 145
methodology, 31, 46
methods, xi, 58, 62, 74, 82, 90, 125, 130,
    144, 145, 148, 149, 169, 180, 203, 205

Index 217

microbial, 76, 78, 104, 188, 189
microorganisms, 181, 186
micropore, 192
microporous, 183, 184, 193, 203, 204, 206
microwave, 144, 147, 148, 152, 165
migration, 147
minerals, 28, 43, 76, 191, 193, 200, 202
minor, 95
misleading, 116
mixing height, 92
mixture, 48, 51, 54, 57, 59, 62, 63, 64, 67, 70, 71, 119, 123, 146, 148
mobile sources, 85, 173
model, viii, x, 2, 3, 4, 7, 8, 9, 10, 13, 15, 16, 23, 25, 26, 85, 86, 87, 88, 89, 90, 91, 93, 96, 98, 104, 126, 128, 129, 131, 136, 193, 197, 201, 207
modelling, viii, ix, 2, 3, 7, 24, 26, 74, 80, 87, 90, 93, 133, 205
modification, 42, 180
moisture, 44, 186, 190
molecular formula, 62, 70, 105
molecules, 51, 52, 62, 70, 105, 108, 109, 145, 190, 192
monomolecular, 192
monoterpenic hydrocarbons, 51
morbidity, 113
morphology, 24, 156
mortality, 97, 101, 113, 134, 177
myrcene, 50, 51, 53, 56, 58, 62, 64, 71, 106

N

nanocrystalline, 153
nanoparticles, 146, 147, 148, 150, 152, 158, 164, 165, 166
naphthalene, 33
native species, x, 104, 115, 116, 117, 119, 122, 123, 124, 125, 129
native trees, x, 104, 119, 130
natives, 115
natural clays, 171, 191, 192
natural conditions, 51, 57, 62, 68, 77
natural emissions, 58, 67
natural environment, 43, 44, 58, 130

natural process, 4
natural sources, vii, 80, 104
natural zeolites, 184
needles, 30, 35, 43, 44, 47, 49, 57, 58, 59, 62, 64, 65, 69, 71, 76, 78, 108, 138
neutral, 4, 6, 7, 8, 11, 12, 24
neutral atmospheric boundary, 11
neutral conditions, 6
neutral stability conditions, 7, 8, 11, 12
neutral thermal stratification, 6
nitrogen oxides, ix, x, 28, 79, 80, 103, 110
nitrogen-free hydrocarbons, 42
non-destructive, 181, 183, 185
non-dimensional standard deviations, 13
non-methane, 57, 77, 81, 97, 100, 104, 200
non-monotonic, 48
non-terpenic, 33
non-terpenoid BVOCs, 67

O

observations, 91, 139
occurrence, 36, 42
oceans, 104, 173
oils, 36, 38, 76, 104, 105
opportunities, 125, 191
optimisation, 44
organic acids, 33, 141, 173, 186
organic carbon, 95, 97, 98
organic chemicals, 28, 171
organic species, 109
organisms, 105, 106
overestimation, 129
oxidation reactions, 161, 162
oxidative process, 186
oxide, 109, 144, 146, 150, 151, 152, 156, 157, 159, 161, 162, 165, 179, 200
oxygen, x, 30, 38, 103, 108, 113, 146, 149, 151, 152, 154, 161, 162, 166, 172, 178, 185, 190
oxygenated compounds, 28
oxygenated species, 81
ozone layer, vii, 170, 178
ozone-precursor, 82

P

paints, 105, 172, 176, 204
parameters, x, 7, 11, 34, 44, 47, 48, 82, 90, 91, 92, 104, 189, 190, 198
parks, 86, 89, 91, 105
particles, 9, 28, 29, 38, 43, 108, 147, 149, 150, 151, 152, 153, 154, 156, 157, 159, 179
particulate, x, 28, 103, 113, 130, 186, 190
pathogens, 105, 108, 189
peak, 53, 55, 65, 72, 73, 84
peak levels, 84
performance, xi, 144, 146, 152, 153, 158, 161, 162, 166, 171, 180, 183, 189, 190, 197
periods, 44, 83, 125
peroxy radicals, 109
peroxyacetyl nitrate, 112
pesticides, 172
petrochemical, 104, 173, 176
pharmaceutical, 182
phenol, 33, 57, 61
phenol derivatives, 33
photochemical, vii, ix, x, 29, 43, 80, 81, 82, 85, 86, 87, 93, 94, 96, 99, 100, 104, 109, 114, 144, 145, 170, 171, 177, 179
photochemical models, 29, 94, 100
photochemical oxidants, 81, 93
photochemical oxidation, ix, 80, 81
photochemical pollution, 43, 85, 94, 170, 177
photochemical products, 86
photochemical reactions, vii, 86, 109, 170, 171
photochemical reactivity, x, 81, 82, 100, 104
photochemistry, 86
photodissociation, 108
photoelectron, 144, 149, 166
photosynthesis, 105, 113, 114, 130, 139
phototropic, 106
physical modelling, 2, 3, 7, 24, 74
physicochemical, xi, 105, 144, 171
physiological, 105, 108, 113, 126

phyto-sanitary risks, 43
pigments, 105
pines, 94
plant characteristics, 58
plant species, 32, 43, 44, 58, 60, 74, 77, 90, 94, 106, 126
plants, viii, 2, 32, 34, 42, 43, 44, 58, 62, 76, 94, 105, 106, 107, 108, 114, 132, 136, 141, 178
plasma, 190
plasma technology, 190
plastids, 106
plume axis, 6, 14
poles, 18, 19
pollinators, 105
pollution, vii, ix, x, 42, 43, 75, 79, 80, 85, 94, 97, 113, 125, 134, 140, 141, 143, 145, 146, 163, 170, 171, 177, 180, 188
polycyclic aromatic hydrocarbons, 43
polygons, 91
polymers, 183, 184
porosity, vii, 1, 8, 191
porous environment, viii, 2
potential, ix, x, 28, 29, 37, 39, 75, 80, 83, 85, 98, 103, 112, 114, 115, 133, 136, 145, 164, 191, 196, 197, 200
power, 11, 44
power law exponent, 11
precursors, 34, 80, 96, 100, 150, 177
predators, 105
pressure, 4, 9, 31, 32, 166, 171, 172, 183, 188, 190, 192, 202, 206
pressure distribution, 9
pre-treatment, 151, 152, 153, 158, 161
prevention, 32, 36
preventive, 180
processes, vii, 4, 26, 42, 81, 82, 87, 105, 146, 150, 171, 173, 178, 180, 182, 183, 185, 186, 187, 189, 190, 191, 195, 203, 204
production, x, 44, 62, 81, 93, 101, 103, 173, 174, 179, 184
productivity, 113, 114
program, 31, 91
project, viii, 1, 2, 3, 4, 8, 23, 97, 131

Index 219

propagation, 42, 89
propane, 87, 146, 151, 152, 154, 160, 161, 164, 165, 166
propene, 146, 161, 165
properties, viii, xi, 2, 3, 7, 8, 9, 12, 13, 23, 76, 144, 146, 149, 153, 162, 171, 172, 190, 191, 197, 200, 201, 202
proportions, 54
protocols, 87, 170, 199
purposes, x, 16, 104, 116, 126, 197
pyrolysis, 42, 47, 48, 74, 75, 76, 78
pyrolysis phase, 47
pyrroles, 33

Q

qualities, 130
quantification, viii, 27, 29, 31, 46, 47, 75, 82, 87, 90, 115
quantity, 44, 50, 51, 52, 53

R

radiant panel, 44, 45, 50, 53, 74
radiant panel heat flux, 50, 53
radiation, 91, 107, 115, 133, 178, 189
radicals, 81, 92, 109, 111, 112, 130
ranges, 8, 11, 13, 89
rate, 4, 11, 42, 43, 45, 47, 48, 74, 75, 85, 87, 95, 105, 126, 139, 152, 159, 162, 173, 180, 182, 186, 187, 190, 205
reactants, 92
reaction, 87, 92, 108, 109, 110, 111, 130, 136, 147, 151, 152, 153, 154, 156, 159, 161, 162, 164, 178, 179, 185, 186, 190, 203
reactive, 2, 42, 81, 87, 96, 109, 113, 130, 178, 190
reactivity, x, 81, 83, 85, 86, 87, 100, 101, 104, 106, 114, 137, 146, 161, 197
reactor, 78, 182, 184
reality, 116
recovery, 43, 87, 180, 181, 182, 185, 188, 191, 197, 203

recuperative, 171, 185
redox properties, 146
redox species, 161, 162
reducibility, 161
reducible, 161
reduction, xi, 4, 43, 52, 75, 113, 129, 144, 145, 147, 148, 151, 161, 165, 171, 179, 180
reforming, 178
refractory, 153
refrigerants, 178
refrigeration, 182
regeneration, 43, 81, 93, 156, 183, 184, 185, 195, 197, 202, 203
regenerative, 186
regional scale, 85, 94
registers, 174
regulations, ix, 80, 170, 180
reliability, 92
remediation, xi, 170
remote sensing, 74, 82, 90, 91, 92, 95, 100, 101
repeatability, 47
repercussions, 177
reproducibility, 3
reproduction, 43
reproductive, 177
research, vii, 4, 7, 23, 130, 131, 133, 134, 163, 184, 190
residential, 37, 82, 91, 176
resins, 105
resistance, viii, 2, 3, 147
resolution, 10, 23, 90, 91, 92, 93, 144, 147
respiratory, 29, 42, 75, 113, 177
retardation, 113
retention, 31, 43, 46, 47, 50, 54, 56, 59, 60, 66, 129, 130
retention times, 31, 46, 47, 54, 59
reusable materials, 182
reversibility, 171
Reynolds number, 7
rings, viii, 2, 3, 8, 10, 148
risk, ix, 28, 29, 42, 177, 178
roughness, 9, 10, 11, 12, 14, 23, 25, 130
roughness length, 11, 12

220 Index

rural, 81, 82, 83, 84, 93, 100, 141, 173
rural sites, 93, 100

S

samples, 5, 6, 9, 47, 70, 82, 83, 84, 85, 86, 93, 97, 154, 156, 192
satellite images, 90, 91, 92
saturation, 52
scale, vii, viii, 1, 2, 3, 7, 9, 10, 11, 12, 13, 16, 17, 24, 25, 44, 47, 48, 62, 74, 76, 77, 86, 87, 93, 177
scheme, 4, 6, 172
scrubbers, 181
seasons, x, 32, 35, 36, 42, 81, 93, 104, 115, 116
sectors, 172, 174, 176
sediments, 104
selectivity, xi, 144, 146, 149
sensitivity, 93, 113, 144, 156, 157, 162
separation membranes, 181
sesquiterpenes, 34, 55, 65, 67, 68, 81, 107
sesquiterpenes compounds, 34
shear stresses, 11, 14
shrubland, 91
shrubs, 44, 70
significant, 4, 28, 43, 51, 53, 55, 81, 83, 84, 87, 89, 91, 92, 93, 95, 105, 113, 116, 117, 119, 125, 130, 132, 150, 151, 158, 173, 195, 196, 197
silicalite, 193, 195
silicates, 192, 195
similarities, 3, 126
singulet oxygen, 108
sinter-ability, 153
skeletal formulas, 54, 59
smog, vii, x, 144, 145, 170, 177, 179
smoke, viii, 27, 28, 29, 36, 38, 39
soil, 28, 43, 68, 77, 78, 129, 192, 206
soil characteristics, 68
soil erosion, 28
solids, 148, 193, 195, 201
solubility, 105
solvents, vii, 105, 170, 173, 174, 176, 177, 178, 204, 206

solvothermal method, 147, 148
solvothermal reactions, 147, 148
sorbent, 45, 46, 183
source, viii, 1, 2, 4, 5, 9, 11, 13, 14, 16, 20, 25, 31, 46, 81, 82, 83, 86, 87, 88, 89, 94, 95, 98, 99, 100, 105, 106, 145, 146, 171, 173, 180, 186, 187, 205
species emissions, 68
specificity, 94
spices, 105
spinels, 153, 166
spring, x, 83, 93, 104, 115, 116, 122, 123, 125, 126
stability, xi, 7, 8, 11, 12, 24, 144, 146, 147, 152, 153, 154, 162, 166
stabilization, 146
stages, x, 36, 104, 115, 116, 140
standard deviation, 4, 6, 13, 15, 35, 116, 117, 118, 119, 122, 126
standardization, 115, 116, 125
stationary sources, 100, 145, 173
statistical, 3, 20, 90, 113, 116, 117, 125
statistical methods, 90
stickiness, 190
stoichiometry, 152
storage, 163, 173
strategy, x, 3, 7, 8, 82, 103, 125
stream, 180, 181, 182, 183, 185, 186, 190
structure, xi, 6, 12, 24, 25, 26, 107, 144, 146, 153, 154, 156, 157, 161, 162, 166, 183, 184, 186, 192, 197, 206
Styrofoam, 8
subsequent, xi, 81, 82, 93, 169
substance, 52
substitution, 180
subtropical, 81, 92, 100
sub-urban, 83, 95
Sulphur hexafluoride (SF6), 4
Summer, 6
supercapacitors, 145
surface layer, 11
symptoms, 177
synthesis, 3, 39, 105, 145, 147, 148, 163, 164, 166

Index

221

systems, xi, 33, 131, 144, 145, 146, 147, 177, 184, 185, 186, 197, 206

T

tangled plastic, 8

tank, 185

taxonomical, x, 104, 126

technique, 4, 5, 47, 58, 62, 90, 115, 138, 180, 181, 182, 185, 186, 189, 190

technologies, xi, 169, 170, 171, 180, 181, 186, 187, 188, 190, 191, 206

temperature gradient, 4, 6

temporal variation, 80, 82, 91, 100

tendency, 51, 60, 62, 148

terpene emissions, 68

terpenes, 34, 51, 77, 106, 107, 109, 111, 114, 116, 130, 131, 133

terpenic compounds, 33, 53

terpenoid compounds, ix, 41, 74

terpenoids, 62, 70

tests, viii, 2, 15, 50, 157, 158, 161

thermal catalysis, xi, 169

thermal degradation, 51, 53, 62, 65, 70, 74

thermal effects, 7

thermal oxidation, 185, 186

thermal power, 44

thermal properties, 7

thermal stability, 147, 166

thermal treatment, 158, 162

thinners, vii, 105, 170

threshold temperature, 53

time-averaged, 13

tolerance threshold, 9

toluene, 81, 83, 85, 87, 98, 100, 146, 165, 172, 199, 200, 203, 204, 205

topography, 81

tower, 4, 5, 6, 12-16, 18, 19, 21, 23

toxic, xi, 4, 112, 169, 189

toxicological studies, 80

trace minerals, 28

tracer, viii, 1, 4, 5, 6, 7, 9, 13, 16, 17, 23, 25, 99, 138

tracer experiments, 4, 6, 9, 23

Tracer-gas experiments, viii, 2

traffic emissions, 84

transfer, 46, 47, 183

transition, x, xi, 13, 144, 145

transmission, 144, 147

transport, vii, 1, 2, 3, 8, 23, 51, 52, 62, 65, 70, 81, 86, 87, 93, 98, 137, 173, 174, 176, 198

transportation, 173

travel altitude, 20, 21

travel distance, 20

travel time, 20, 22, 23

treatment, xi, 147, 151, 152, 156, 158, 161, 162, 167, 169, 171, 180, 182, 183, 184, 185, 186, 187, 188, 190, 191, 199, 201, 202, 206

tree, 4, 7, 8, 10, 12, 16, 20, 88, 89, 90, 106, 114, 115, 117, 119, 121, 123, 126, 129

tree height, 4, 7, 8, 10, 12, 16, 20

tree species, 4, 88, 89, 90, 114, 115, 123, 126, 129

triplet oxygen, 108

tropical, 39, 87, 89, 93, 94, 101, 136, 137, 138

troposphere, vii, 1, 78, 108, 109, 111, 178, 179, 202

tropospheric, ix, 42, 79, 80, 81, 92, 95, 109, 111, 113, 114, 130, 133, 134, 138, 140, 177

turbulence, viii, 2, 6, 9, 11, 12, 13, 16, 24, 25

turbulence intensity profiles, 11

turbulence profiles, 12

turbulence structure, 6

turbulent dispersion process, vii, 1

turbulent flow, 3, 12, 25

turbulent fluctuations, 11, 20, 21

turbulent fluxes, 3

turbulent velocity, 8, 23

turbulent velocity profiles, 8, 23

U

unstable, xi, 6, 111, 144, 178

unstable conditions, 6

urban areas, ix, 80, 83, 88, 129, 173, 174

urban forest, x, 104, 108, 115, 121, 140
urban trees, 115, 118, 119, 120, 121, 122, 123, 124, 125, 129

V

vapour pressure, 4
vapour-liquid, 62, 70
variability, 90
variables, 107, 108, 116
variation, 53, 82, 90, 91, 93, 138
variety, 87, 130, 145, 170, 172, 177, 183
various, vii, ix, xi, 43, 76, 80, 104, 105, 108, 113, 114, 131, 140, 144, 145, 146, 151, 152, 154, 172, 173, 180, 182, 184, 187, 192, 206
vegetal fuels, 74
vegetal species, ix, 41, 42, 43, 47, 48, 51, 59, 70, 72, 73, 74
vegetation, ix, 3, 7, 8, 24, 25, 29, 32, 33, 37, 39, 41, 42, 43, 50, 57, 75, 76, 77, 78, 80, 81, 82, 83, 89, 91, 92, 93, 94, 104, 105, 126, 129, 130, 133, 134, 135, 139, 142, 174, 179
vehicles, 105, 176
vehicular emission, 83, 84, 93
vehicular markers, 83
velocity, 3, 4, 7, 8, 9, 11-16, 23, 185
velocity fluctuations, 12
vertical distribution, 12, 13
vertical fluctuations, 12
vertical gradient, 13
vertical layers, 16, 20
vertical profile, 16, 20
vertical turbulent fluctuations, 20, 21
vessel, 148
visible, 113
VOCs emissions, vii, 1, 35, 36, 62, 65, 81, 173
volatility, 105, 106, 190
volcanoes, 104, 173
volume, 9, 11, 13, 26, 44, 192, 206
volume flow, 13

volumetric flow, 182

W

water, 28, 43, 51, 53, 65, 70, 78, 105, 108, 154, 181, 184, 185, 186, 188, 190, 197, 200, 202
water evaporation, 51, 53, 70
water infiltration, 43
water vapor, 28, 108
waxes, 105
weather conditions, 28
wetlands, 32, 173
wetness, 150, 151, 152, 154
wildfires, 29, 32, 37
wildland fires, viii, 27, 28, 39, 42
wind direction, 4, 6, 7, 12, 14
wind forces, 7, 24
wind gusts and sweeps, 8
wind profile, 8, 11, 12
wind speed, 4, 6, 92
wind tunnel, vii, viii, 1, 2, 3, 4, 6, 7, 8, 9, 10, 11, 12, 13, 14, 15, 17, 23, 24, 26, 74
wind tunnel experiments, 10, 12, 13, 14, 23

X

Xylene, 194

Y

yield, 92, 113
young species, 117, 119

Z

zeolites, 183, 184, 188, 193, 195, 198, 199, 202, 203, 205, 206
zones, 187